CREATURES OF REASON

CREATURES OF REASON

JOHN HERSCHEL
AND THE INVENTION OF SCIENCE

STEPHEN CASE

UNIVERSITY OF PITTSBURGH PRESS

Published by the University of Pittsburgh Press, Pittsburgh, Pa., 15260
Copyright © 2024, University of Pittsburgh Press
All rights reserved
Manufactured in the United States of America
Printed on acid-free paper
10 9 8 7 6 5 4 3 2 1

Cataloging-in-Publication data is available from the Library of Congress

ISBN 13: 978-0-8229-4838-4
ISBN 10: 0-8229-4838-9

Cover art: Sir John Frederick William Herschel. Mezzotint by W. Ward, 1835, after
H. W. Pickersgill. Courtesy of Wellcome Library / Wellcome Images, London.
Cover design: Melissa Dias-Mandoly

For my mom and dad, Laurette and Bill Case

CONTENTS

Acknowledgments   ix

Introduction   3

Prologue: The Astronomer and the Autocrat   11

1. A Cambridge Discontent   19

2. A Branch of Elegant Literature   44

3. Revolt of the Astronomers   71

4. Valleys and Summits   96

5. Grand Tour   119

6. A London Discontent   145

7. Stability of the System   165

8. Reform Defeated   185

9. The Invention of Science   208

Epilogue   229

Notes   233

Bibliography   253

Index   265

ACKNOWLEDGMENTS

I owe a debt of gratitude to many for the support, encouragement, and guidance that made this project possible as it developed. Special thanks are due to Michael Crowe, who first introduced me—as a graduate student newly arrived from the sciences to the humanities—to a character named John Frederick William Herschel. Michael's scholarship and the work he and his team invested in *A Calendar of the Correspondence of Sir John Herschel* (Cambridge University Press, 1998) created what remains the essential entrance into the immense mass of Herschel materials for any Herschel scholar. I have also benefited from conversations with Gregory Good, Simon Schaffer, Woody Sullivan, Marvin Bolt, Larry Schaaf, Kelley Wilder, and the late Michael Hoskin, who cheerfully shared his enthusiasm and wealth of knowledge on life at Cambridge. I also thank the participants of a panel session on John Herschel at the History of Science Society meeting in 2019 in Utrecht, who offered useful thoughts and feedback on this project in an early form, and to James Secord, whose work besides helping situate Herschel in the publishing world provided an ideal illustration of the nineteenth-century British view of stable society (reproduced in chapter 2).

I am grateful to Will Herschel-Shorland, curator of the Herschel Family Archive, for his support and for permission to quote from John Herschel's unpublished correspondence, manuscripts, notebooks, and journals, as well as to the Royal Society, the Royal Astronomical Society, the Harry Ransom Center at the University of Texas, Austin, and the Syndics of the University of Cambridge for permission to quote from materials held in their collections. Aaron Pratt at the Harry Ransom Center and Kathryn McKee at St John's College Library provided helpful assistance regarding materials in their collections. Jacklyn Burns at the J. Paul Getty Museum, Los Angeles, and Ariana Torres at the Museum of Photographic Arts, San Diego, helped in obtaining reproductions of the Herschel camera lucida drawings held at their museums. Graham Nash and his agent Mark Spector also provided guidance in locating some of these images.

I am extremely grateful to Sandy Harris, interlibrary loan librarian extraordinaire at Olivet Nazarene University as well as the entire Benner Library staff, and to my provost, Stephen Lowe, for continued support and encouragement in multiple ways throughout the years. Portions of this project were supported by the Craighton T. and Linda G. Hippenhammer grant for faculty research at Olivet Nazarene University as well as a scholar-in-aid award from the American Institute of Physics. My friend and colleague Christopher White provided thoughts, perspectives, and encouragement that were invaluable throughout the entire writing process. A faculty writing group at Olivet helped see this project through its earliest phases and included, among others, Nancy Bonilla, Justin Brown, Paul Kenyon, Tiffany Greer, Brian Stipp, and Derek Rosenberger. Thanks to Stefan and Ari Frunze of Stefari Café in Kankakee for all the coffee.

Two anonymous reviewers offered insightful comments and suggestions that strengthened this work, though of course any errors that remain are my own. I'm thankful to Abby Collier and her entire team at University of Pittsburgh Press for making this book a reality. And finally, I am continually grateful to my wife, Christine, and my children for the years of patiently sharing me with this project.

CREATURES OF REASON

INTRODUCTION

John Herschel (1792–1871) was angry. The young natural philosopher, son of the world's most famous astronomer, top of his class at Cambridge University, and rising star among the scientific elite of Regency London, was fed up with the state of mathematics in Britain in general and as taught and practiced at Cambridge in particular. His frustration had been simmering for years as he worked toward mathematical reform through his own research and publications, but in 1816 the publication of a second edition of *Principles of Fluxions* by the venerable Cambridge don William Dealtry (1775–1847) was the last straw.

It was not simply that Herschel had hoped his own textbook, published two years earlier, had made works like Dealtry's irrelevant. It was not that Dealtry's book was bad and contained, according to Herschel, "nothing like uniformity of method, no pervading principle." It was not even that the book was riddled with errors. For Herschel, the problem of Dealtry's work was deeper and more fundamental. By posing as a treatise written to "instruct the rising generation in the principles of the differential calculus—to teach the young philosopher the nature and use of that all powerful instrument of research which is to be his guide through every intricacy of nature," the book was actively harmful to students and to the British scientific endeavor.

Herschel recognized that advanced mathematics had become indispensable for investigating the physical world. Mathematics had been central to natural philosophy for centuries, but new tools and methods developed throughout the 1700s had pushed this relationship to deeper levels and higher complexity—and Britain was being left behind. Though England proudly claimed the heritage of Isaac Newton (1643–

1727), works like Dealtry's showed how badly the nation was neglecting the expansions, transformations, and developments of Newton's work being made on the Continent, especially in France. Dealtry's book was backward and antiquated, according to Herschel, ignoring the new approaches to calculus and algebra (known together during this period as *analysis*) that Herschel felt were necessary for natural philosophy to make true progress.

British mathematicians remained stuck in the past, using outdated formalism and notation. And the success of Dealtry's book—evidenced by the appearance of a second edition—dashed Herschel's "aspirations after better things" and convinced him "that the period is yet far distant when mathematical science in the state of perfection to which our continental neighbours have brought it, will cease to be exotic to the British soil." In other words, Britain still had a long way to go. Herschel believed he was fighting for the future of the scientific endeavor, and pedantic texts like Dealtry's that centered on disconnected series of mathematical applications were worse than useless. They were actively harmful. Their "cloud of consecrated puerilities," Herschel railed, kept students and readers "distracted by an impertinent detail of trivial applications" and completely missed the point of mathematical instruction.[1]

Herschel was in a position to judge. Though known today primarily to historians of science and photography, Herschel in his lifetime became Britain's best-known natural philosopher, a world celebrity and scientific polymath of the generation in which the term *scientist* was itself invented. The son of William Herschel (1738–1822), discoverer of Uranus and constructor of the world's largest telescopes, Herschel sailed through Cambridge taking highest honors, conducted groundbreaking work in chemistry and optics (as a photographic pioneer he discovered the fixing agent used by traditional photography for more than a century and coined the terms *negative* and *snapshot*), helped establish a mathematical revolution, extended his father's astronomical surveys to the entire sky, and wrote the popular texts by which a generation of readers learned what it meant to "do science." Along the way, John Herschel established the practices of this new approach to natural philosophy in a period where traditional, hierarchical approaches to knowledge were giving way to new forms based on gentility, dispassion, and social credibility. The practice of science, under Herschel's influence, was becoming highly organized yet individualistic, culturally relevant and government funded but free from political control, and abstractly mathematical while grounded on observation. In sum, Herschel helped give to natural philosophy the contours of modern science. This culminated in his publication in 1831 of the book that turned a generation of natural philosophers into scientists: *A Preliminary Discourse on the Study of Natural Philosophy*.

In his *Discourse*, Herschel offered an appeal for and outline of this new scientific life. Among other things, he provided a definition of scientific theories as "creatures of reason rather than of sense," of creations of thought built from natural laws just as laws in turn were built from sense observations of the physical world.[2] But on another level, the term *creatures of reason* could also refer to a new type of scientific practitioner: the natural philosopher beginning to transition into the modern scientist. Finally, in a deeper sense, Herschel had created himself to be an ideal "creature of reason," and his entire early career was the story of this self-fashioning, from his initial frustrated mathematical reforms to becoming the central figure of the nineteenth-century British scientific community.

The *Preliminary Discourse* has long been acknowledged as a seminal text in the history of science. It was critical in the transition from natural philosophy to modern science that took place in the 1830s with the beginnings of professionalization of science into disciplines, the spread of scientific ideals to the middle classes through popular literature, and the application of scientific principles to society. For philosophers of science, the *Discourse* offered the first modern treatment of the inductive method. The work, in which Herschel set out his rules of scientific reasoning and theory formation, has been taken as offering everything from a codification of induction to an appeal to radical hypothesis to the first account of epistemological surprise as sign of robust theory formation. The *Discourse* was important for initiating one of the major philosophical debates of the nineteenth century between John Stuart Mill (1806–1873), who argued that experience was all-important in forming knowledge, and William Whewell (1794–1866), who argued that discoveries were made by the structures of human thought.[3]

Historian James Secord has argued that the majority of contemporary readers of the *Discourse* would have read it not as a philosophical text but, considering the book's price and format when originally published, as a conduct manual, a guide for the rising middle class to accrue social credibility by learning to think and act more scientifically. In this light, the *Discourse* was a product of the London publishing industry, shaped by market forces to appeal to a specific audience. As the steam press reduced the cost of printed books and politicians built programs of social reform on educating the populace through literature, the *Discourse* was a product capitalizing on a growing market, and Herschel was the scientific star pegged to write what would become one of the first science best sellers. As such, the work had an undeniable inspirational influence, the most famous example being the role it played for Charles Darwin (1809–1882) in providing not only a model for scientific theory formation but for a scientific life.[4]

The *Discourse* is thus a central text to understanding the origins of modern science. But what of the origins of the *Discourse*? The immediate context of the *Preliminary Discourse* is its author's life: not only how Herschel came to write it but how he became the person to write it—the experiences, values, and influences that shaped him as a creature of reason from his childhood through his early career to become one of the most recognized scientific authorities of his age. Placing the *Discourse* in this narrative allows new aspects of the work to emerge, in particular how the *Discourse* formed a synthesis of Herschel's previous work, built on his experiences at the telescope, in the laboratory, and on the mountaintop.

A biographical context also reveals the *Discourse* as a *reforming* work. A theme through Herschel's early life was his desire to see natural philosophy pulled from its traditional moorings in outdated mathematical formalisms and elite institutions to become more dynamic, efficient, and meritorious. Herschel found his attempts in this direction continually stymied throughout his years in Cambridge and then London, even as Tory conservatives in Parliament resisted political reform on the national level. When marketing and publishing forces offered Herschel the opportunity to write a popular treatise on natural philosophy, he used this chance to bring his reforming view of science to the public at large. This served the rationalizing social ambitions of the period as well as provided a rebuke to partisans (including some of his closest friends) arguing questions of privilege and financial support within scientific society. The virtues of Herschel's ideal natural philosopher—patience, disinterest, dispassion—were his own lived ideals, not the bickering controversies engulfing the Royal Society, which was supposed to be the nation's leading scientific institution. Herschel has often been seen as content to avoid controversy and work within established institutions. In the context of his life and career, however, the *Discourse* became his most effective instrument of reform. By the close of 1830, after only a few breakneck months of writing, Herschel created what would become the definitive treatment of the scientific endeavor for a generation. Herschel set out to reform natural philosophy; along the way, he helped invent modern science.

Though written almost two centuries ago, at the close of the Romantic era, Herschel's vision of science offered in the *Discourse* is of enduring importance in the twenty-first century. The *Preliminary Discourse* defined the nature of scientific practice for the age in which science became modern: professionalized, divided into disciplines, and inextricably linked with mathematics. During Herschel's early career, the increasingly complex mathematics required for natural philosophy created a perceived hierarchy between individuals who were trained in those mathematical methods and "amateur" natural philosophers who made observations, a division within science that remains in some form today. Science, it is claimed, is open to all, yet to produce scientific theories—not simply observations—mastery of advanced mathematical methods is required. Many can grapple with the concepts of physics, for instance, but only a mathematical physicist can use the observations from particle colliders to create new theories. Science is a plutocracy, and mathematics remains the currency of the realm.

Rightly or wrongly, Herschel was among the first of his generation to make this distinction. Even as he worried Britain had lost the mathematical edge it needed to advance in science, he was concerned that this mathematical necessity would undermine the reforming, democratic aspects of the natural philosophical endeavor. Herschel wrote in a period in which society was beginning to be modeled along scientific principles. As industry, commerce, wealth, and literacy grew, so did the need for an understanding of the methods and practices of science. A good citizen was someone who understood how to think scientifically. But was this even possible if only those trained in mathematical analysis could create the products of natural philosophy? Herschel wrestled with these issues, and the answers he forged in his *Preliminary Discourse* have shaped dialogue and conceptions of science in society ever since.

Since at least the 1960s, traditional views of the history of science have placed the origins of modern science within the Scientific Revolution, exemplified in the application of mathematics to the study of nature in the works of Galileo Galilei (1564–1642) and Newton and the inductive processes of Francis Bacon (1561–1626). More recently historians have argued that this misconstrues these early methods of understanding the world and that prior to the late 1700s and early 1800s this work should be understood in terms the practitioners themselves used, as *natural philosophy*, a mode of thought and investigation quite different from what we would identify as science today. As William Lubenow emphasizes in his study on learned societies of the nineteenth century, during this earlier period there was quite simply "no such thing as 'science.'"[5]

It was only the 1830s that saw the invention of science as a new and distinct undertaking. Natural philosophy was the effort to understand the world as created by God, and its virtues were largely confined to the gentlemanly elite. It was a unified endeavor, encompassing the entire natural world. In the period of Herschel's *Discourse*, however, the virtues of natural philosophy were expanded to all classes. In Britain, new professional opportunities arose, allowing the chance to pursue study of the natural world independent of the established church or social position. Optics, electromagnetism, chemistry, mineralogy, and geology flourished, giving rise to new scientific disciplines. The process of investigating the natural world went hand-in-hand with the growth of empire as science became globalized, and a new generation of philosophers began to critically examine what natural science was and how it worked, even as the industrial revolution transformed society. Herschel and the *Preliminary Discourse* were central to this transition. Understanding the context of this work thus illuminates this transformation from natural philosophy to science.[6]

Herschel has been referred to as "the forgotten philosopher." No theories are attributed to him, though his work showed how scientific theorizing was done. No discovery or invention bears his name, though his fingerprints are everywhere, from the vocabulary of photography to the names of moons of the outer solar system to the general deep-sky catalog used by astronomers today. Often lost in his father's shadow, the son nonetheless made sidereal astronomy mainstream. Herschel lies buried in Westminster Abby near Newton (and beside Darwin), a testimony to his place at the center of the scientific community and a reminder of how much we understand as the nature of modern science was established by Herschel and his *Preliminary Discourse*. Herschel's *Discourse* became, in sum, the *apologia* for a scientific life by the century's foremost natural philosopher. As such, the story of the *Discourse* and its composition is the story of the development of a new kind of scientific practice, a key step in the transition from natural philosophy to modern science.

A focus on the *Preliminary Discourse* allows a biographical approach to Herschel himself. Herschel embodied the scientific ideals of the era, values he articulated in his own persona before writing them into the *Discourse*. Though Herschel features centrally in the historiography of nineteenth-century science, no extensive biography of his life has been written. One reason is the daunting scale of his long career, which included hundreds of publications; thousands of pieces of surviving correspondence; important contributions to chemistry, mathematics, photography, optics, and astronomy; extensive travels, including his four years at the Cape of Good Hope; and a period holding the important governmental post of master of the mint. Using the *Preliminary Discourse* as a frame for Herschel's early biography provides a useful chronological breakpoint—the book's publication—as well as a natural culmination of the influence and credibility he slowly and steadily built from his student years onward. The story of the *Preliminary Discourse* is thus the story of Herschel's early career, from his matriculation at Cambridge in 1808 to preparations for his departure for the Cape in 1833.[7]

To write the *Discourse*, Herschel had to first become the person who *could* write the *Discourse*. The story of Herschel and the invention of science is thus the story of his invention of himself as the ideal natural philosopher, moving from the shadow of the telescope to becoming a reforming force in British society, drawing on his travels lost on an Alpine glacier in the middle of the night or scaling the slopes of Mount Etna, the influence of his famous father and aunt, and ultimately his relationship and marriage to Margaret Brodie Stewart (1810–1884), the daughter of an evangelical minister. It is a story of his colleagues, the pugnacious and sometimes infuriating father of computing, Charles Babbage (1791–1871), and the idealistic barrister James Grahame (1790–1842), and of middle-class accountants threatening the hegemony of aristocratic privilege. And it is ultimately the story of Herschel's failure at the ballot box but success through a runaway best seller that took the reform of science to the streets, making the practice of the halls of Somerset House the thought and conversation of the working class.

Herschel did not simply write about natural philosophy; he was the Romantic ideal of its practice. Herschel's vision of science in the *Discourse* was successful because the scientific ideal he laid out within was one he steadfastly pursued in his own life. His early career is the story of a polymath with one foot in the traditional world of natural philosophy and the other in the new realm of modern science, but it is also the account of a young person trying to live up to a very large name in a changing world.

Fortunately for Dealtry and his mathematics textbook, Herschel never published his damning review. It exists only in a manuscript draft, at the top of which Herschel at some point added a begrudging note that though not printed, "it ought to have been. It is very severe but perfectly just & would have done much good." At the time, Herschel was a fellow of St John's College and thus part of the institution he was railing against. Throughout his life, Herschel favored gradual reform, politically, socially, and pedagogically. His unpublished review gives a rare glimpse of the passions below his usually equanimous surface. "The force and vigour of the Student's mind is squandered away," Herschel thundered, "his spirit of enquiry quashed, and his relish for mathematical speculations in general, destroyed, in running this unmeaning round which leads to nothing; while the vast expanse of attainable knowledge is studiously kept concealed from his view."

Though Dealtry and Cambridge were spared the published ravages of his pen, Herschel worked to open that "vast expanse" to students and society by effecting the changes he wished to see. His reform of science would begin with a mathematical assault on the conservative courts of the university. The esteem in which Herschel held the new mathematical developments coming out of France—against which the traditional mathematics of Dealtry compared so poorly—went back to Herschel's childhood. William Herschel had made sure his son's education included exposure to the new mathematical analysis, but the roots went even further than that, to when John accompanied his father to Paris as a child and there got his first glimpse of a new scientific world. For William, the Paris trip marked the culmination of his own rise from itinerate immigrant musician to renowned natural philosopher, and for young John, that trip would have a lasting influence.

PROLOGUE

THE ASTRONOMER AND THE AUTOCRAT

When John Herschel was ten years old, peace broke out in Europe. Since the French Revolution, Britain and its network of European allies had been at war with the armies of the new French Republic. For most of John's life, the French had soundly defeated the allied forces on land, conquering portions of Italy, southern Germany, and the Low Countries, in part due to the brilliant military campaigns of Napoleon Bonaparte (1769–1821). In Britain any initial sympathy for the revolutionaries had been quickly quelled by growing horror as France moved toward regicide, authoritative rule, and wars that seemed aimed to export their revolutionary ideals abroad. As the French armies triumphed, it looked to many observers that equality and liberty were on their way to toppling the conservative forces of monarchy and aristocratic privilege. In 1802, however, the conflict ground to a halt long enough for both sides to lick their wounds. The Peace of Amiens meant that for the first time in nearly a decade it was possible for the British to travel freely to Paris, the center of the scientific world.

William Herschel leaped at the opportunity, taking his wife, Mary Pitt Herschel (1750–1832), and their young son, John, with him. William had been to Paris only once before, as an unknown musician in 1772. This time, he would come as a renowned natural philosopher recognized, together with his sister Caroline, as having revolutionized astronomy. Beyond his fame among savants, William was an international celebrity, having done what no one in history had before by discovering an entirely new world in the heavens. Almost exactly eleven years before John was born, William, an immigrant to England from Hannover in what is now Germany, went from being an established musician, composer, and music instructor in the English resort town of Bath to a household name in Britain and abroad with the discovery of the planet that would eventually become known as Uranus. Up until then, the solar system had included five visible planets besides Earth: Mercury, Venus, Mars, Jupiter, and Saturn. Even the telescopic discoveries of Galileo more than a century before had only provided new details of or attendants for bodies already known in the solar system: the surface of the moon, satellites around Jupiter, the phases of Venus. With William's discovery of an entirely new world, the size of the solar system doubled. By the time of the Peace of Amiens, William Herschel was at the height of his fame.

John Herschel grew up in the shadow of his father's astronomical fame. In particular, he grew up in the shadow of the massive forty-foot reflecting telescope, the largest in the world, that his father had constructed outside their home in the village of Slough, near Windsor Castle, after William became personal astronomer to King George III (1738–1820). Though George had lost the British colonies in the New World with the American Revolution, the discovery of a new planet (which William called the Georgium Sidus, or George's Star) may have helped mitigate a measure of British shame at being expelled from North America by a band of rebels. George had lost a continent but gained a world, and he honored its discoverer by making William his personal astronomer and providing him an annual stipend to continue his observations. The massive telescope, and the other reflecting telescopes William constructed with his brother Alexander and exported throughout Europe, represented a new kind of power and prestige. They were the artillery with which to wage a campaign of discovery. William's goal extended far beyond the boundary of the solar system. He wanted telescopes that could peer to the very boundaries of the universe, and the surveys of nebulae, star clusters, and double stars he conducted with them revealed a new vision of the cosmos.

One person in Paris was especially interested in the dynamic universe that William's telescopes revealed: Pierre Simon Laplace (1749–1827), the most powerful and influential natural philosopher in France. Laplace had gained renown in scientific circles before the French Revolution and had the political savvy to survive the following Terror even as friends and colleagues such as the aristocratic chemist Antoine Lavoisier (1743–1794) were sent to the guillotine. In his role at the École Militaire, Laplace had examined and passed Napoleon upon Napoleon's graduation as an artillery officer, and as that officer rose through the ranks Laplace saw his own influence increase, first as minister of the interior and ultimately as president of the French Senate.

Laplace gained fame by applying the new mathematical tools developed in France over the previous century to the physical world. Since Isaac Newton had created calculus to apply the law of universal gravitation to the motion of objects in the solar system, mathematicians such as Leonhard Euler (1707–1783), Jean le Rond d'Alembert (1717–1783), Joseph-Louis Lagrange (1736–1813), and Laplace himself had formalized new more powerful mathematical approaches known collectively as analysis. This development reached its culmination in Laplace's magisterial *Mécanique céleste* (published in five volumes from 1799 to 1825), which offered a comprehensive mathematical explanation of the motion of the planets and completed the project Newton had begun of quantifying the past and future motions of the objects in the solar system. In particular, Laplace's work had bearing on the stability of the cosmos. As Laplace was able to show, variations in the motions of the planets caused by their gravitational influence on each other were periodic within specific boundaries. In other words, the solar system was and would remain stable over long periods of time.

This insight was more than idle scientific speculation. During this period of political transitions, astronomy came to be seen as having important implications for society at large. The revolutions in France and the British colonies of North America raised questions about the assumptions on which a stable society was erected. For thinkers in Britain, who looked askance on the excesses of the French Revolution—the execution of a king, civil unrest, political instability, the overthrow of the aristocracy—astronomy revealed the orderly functioning of a carefully designed universe. Radical upheavals were not part of the natural order, which instead showed a divine and stable hierarchy ordained by a creator.

Order was important in France as well, exhausted as the country was by revolutionary tumults and subsequent European conflict. But in France this was an order founded on mathematical reasoning, a cohesive program of physical research dominated by Laplace, and strong centralized government support. Laplace had become the leader of a school of natural philosophers working to unify scientific knowledge and explain all physical processes in terms of attractive and repulsive forces between particles. His disciples, supported by government funding or lucrative official postings, went on to apply this program to developing an understanding of heat, light, magnetism, and electricity. And because the deposed Bourbon monarchs had shown liberal support of the sciences, the new autocrat Napoleon felt the need to ensure that science flourished even more under his own rule.[1] The scientific scene that William, with John in tow, arrived in Paris to find was one of excitement, optimism, organization, and prestige.

William was quickly immersed in the scientific milieu. In Paris he

13

attended meetings of the first class of the Institut de France, established after the revolution in 1795 to replace the Académie Royale des Sciences and to which he was elected as one of the few foreign members. At the Institut he met figures like René Just Haüy (1743–1822), father of modern crystallography, the astronomer Charles Messier (1730–1817), whose famous catalog of nebulae was a precursor to Herschel's own work, and the mathematician Sylvestre François Lacroix (1765–1843), whose textbooks were promulgating the new French mathematics. Unlike the Royal Society of London, of which William was also a fellow, membership in the Institut was strictly limited to active scientific contributors, with no room for aristocratic dilettantes. It was small, select, and highly professional. Members received government salaries, and the entire body functioned as a formal advisory body to the government on scientific matters. The Royal Society, by contrast, resembled a club for wealthy amateurs. Science in Britain was seen as the concern not of the state but of individuals and voluntary associations. The Royal Society was large and diffuse, with only an informal relationship to the British government. By the end of the eighteenth century, its membership had ballooned to close to five hundred.[2]

During his weeks in Paris, William dined regularly with Laplace. He also met chemist Claude Louis Berthollet (1748–1822), Laplace's colleague and close friend, and traveled to Berthollet's home in the picturesque village of Arcueil. Arcueil, just a few miles outside of Paris, would eventually become an important center for the Laplacian school of physics when Laplace purchased a house on property adjoining Berthollet's. Though John may not have accompanied his father on these visits, the younger Herschel would eventually return to those "avenues, parterres, and lawns, terraces and broad gravel walks" to discuss his father's work and his own contributions to analysis as a mathematician in his own right.[3]

The conversations between William Herschel and Laplace no doubt ranged far and wide, but there was one topic that the doyen of French physical science was especially keen to discuss with the British astronomer: the origin of the solar system. Laplace had developed a theory that the solar system developed from a rotating cloud of gas that collapsed to form the sun. The inclination and direction of the orbits of the planets around the sun (including William's new planet) and the fact that the moons of the planets had similar orbital features pointed, Laplace argued, to a common physical cause. It was statistically unlikely for chance to have arranged all the bodies in such a way. There must be a natural mechanism that could explain these characteristics of the solar system. Laplace's explanation was that if the sun indeed formed of a collapsing, rotating ball of gas, portions of that cloud could have become the planets. These portions would then orbit in the direction the sun rotated and be strung out along the plane of the sun's equator, just as observed. Nebulous, congealing stars like the sun would naturally form systems of orderly planets. Unfortunately, Laplace's theory lacked observational evidence; there was no way to observe this process in action.

By the time of William's visit to Paris, the British astronomer's observations provided hints of support for Laplace's theory. If the universe beyond the solar system was dynamic and changing and if William's observations indeed showed nebulae collapsing to form stars, this would be a powerful corroboration of Laplace's ideas on the origin of the sun and planets. Laplace had provided mathematical reasoning; William's observations could provide evidence of the process of planetary formation in the universe. The critical observation did not come, however, until almost a decade after their meeting: in 1811 William discovered a "nebulous star" that seemed to indicate stars condensed gravitationally from glowing clouds of gas in the heavens. When Laplace read about this observation, he took it as a confirmation of his own theory. Popular writers would ultimately combine William's theory with Laplace's theory to formulate the "nebular hypothesis" as the ultimate origin of planets and stars as well as a symbol of the progressive development of the cosmos.[4]

These developments were in the future, but the work of both Laplace and William Herschel had already begun to bring notions of deep time to the sidereal universe. While in Paris, William met someone doing the same for Earth itself: Georges Cuvier (1769–1832), a young naturalist considered the founder of modern paleontology. Cuvier had quickly risen to prominence in the scientific world with his studies of fossil remains and by the year after Herschel's visit had been named one of the two executive secretaries to the first class of the Institut. This role put him in personal contact with Napoleon, and his growing influence allowed Cuvier to establish a network gathering paper proxies of fossil specimens from collections across Europe. In particular, Cuvier collected evidence and argued in a series of popular works that species existed in the past that no longer existed on Earth; in other words, Cuvier claimed there had been cataclysms in the past that had markedly changed conditions on Earth's surface and driven certain species to extinction. Cuvier's work on fossils opened the door to a history of Earth as extended and dynamic as the celestial history William was building for the heavens.[5]

Much of Cuvier's work would come after William's visit, but while in Paris William likely glimpsed the beginnings of a debate that was to play out in geology, mirroring questions about the nature of the larger universe. In 1801 Jean-Baptiste Lamarck (1744–1829), another French naturalist and colleague of Cuvier, had published his influential work on invertebrates. In the work, Lamarck included an essay on fossils in which he refuted Cuvier's claims and argued instead that fossils were evidence of transmutations or changes in species. Instead of a geological history that included cataclysmic changes, Lamarck argued for a gradualist or even static view in which things ceaselessly and steadily changed but not through cataclysmic revolutions. This view was eventually associated in Britain with the work of James Hutton (1726–1797), a Scottish natural philosopher who had argued something similar for Earth's surface: that the continents were constantly being eroded away while new continents were uplifted from the sea. Both Lamarck in terms of fossils and Hutton in terms of surface processes came to represent a steady-state or even eternal view of Earth's surface opposed to Cuvier's stance. The year of William's trip to Paris, Hutton's views were made more accessible and widely known through the publication of *Illustrations of the Huttonian Theory of the Earth* by another Scottish mathematician, John Playfair (1748–1819). This debate brought Earth itself into questions of whether the universe was eternal or whether it progressed from distinct past states to new states. It was a conversation that John Herschel would find himself engaging when he returned to Paris years later.[6]

In Paris William was exposed to a community of natural philosophers far more organized and supported than in Britain. The learned societies in England were focused primarily on botany, natural history, and agricultural improvements, largely supported by aristocratic interest or patronage and pursued by amateurs haphazardly as their interests dictated or (rarely) by professors at the universities of Cambridge, Oxford, and Edinburgh. In contrast, the natural philosophers at the Institut were, like Laplace and Cuvier, professional civil servants, the most successful of whom were rewarded with government positions and generous salaries. Science in France was seen as a tool of republicanism, equality, humanism, and rationalism, and it was flourishing like never before under the sponsorship of Napoleon, technically still the first consul of France but in practice quickly becoming the nation's all-powerful dictator. Before he returned to England, William would have an opportunity to meet this new patron of science.

On 8 August, a Sunday, William dined with Jean-Antoine Chaptal (1756–1832), a chemist and the minister of the interior, who took the visiting astronomer along with Laplace and a few others to Malmaison, Napoleon's palace home. William was unsure what to expect meeting the man who had conquered much of Europe. The king's astronomer often interacted with the British monarch and his family, but Napoleon represented something new—a self-made ruler, military genius, and dictator. To his surprise, William and his French colleagues found the first consul in his gardens, supervising the digging of an irrigation system. Napoleon, William related in his journal of the trip, was civil and easy to talk with, showing a surprising depth of knowledge. They walked together for about half an hour, with Napoleon politely asking William questions on astronomy. Eventually the first consul invited the group inside to continue the conversation, which ranged from breeding horses in England to the creation of the universe. Napoleon sat and invited William to do the same, but when William noticed that no one else was offered a seat—including the second and third consuls—he remained standing for the entire conversation.[7]

Laplace had recently published the third volume of his *Mécanique*, which he dedicated to Napoleon. The conversation that played out between the mathematician and the dictator at this meeting on the contents of Laplace's work has become the stuff of historical legend. Napoleon is supposed to have remarked to Laplace that there was no mention of God anywhere in Laplace's physical explanation of the planetary system, to which Laplace answered that he had no need of such a hypothesis. Though there is no evidence for this exact exchange, William noted in his journal that Napoleon and Laplace indeed argued about whether the system of the world needed a creator to explain it. When the discussion turned to the extent of the universe, Napoleon asked "in a tone of exclamation or admiration . . . 'and who is the author of all this.'" Laplace, William recounted, tried to show in response that "a chain of natural causes would account for the construction and preservation of the wonderful system." Napoleon was not pleased. "This," the astronomer recorded wryly, "the First Consul rather opposed."

Eventually Napoleon was called to other duties, and his guests excused themselves. William concluded his account of the meeting by reflecting that the truth likely lay between Laplace's and Napoleon's positions: by uniting their arguments, he felt, one could be led to "Nature and Nature's God." William, who had converted from his native Lutheranism when he immigrated to England, had served as organist at the cathedral in Bath, and hosted bishops and clergymen at his home, seems to have been a person of reserved but genuine faith, not an enthusiastic churchgoer but a profound believer in God. Years after his father's death, John would emphasize this faith. "Let it be distinctly understood," he told a Continental correspondent, "that my Father . . . was a sincere believer in and worshipper of a benevolent intelligent and superintending Deity whose glory he conceived himself to be legitimately forwarding by investigating the magnificent structure of the universe." William's astronomy—in contrast with much of the consciously secular philosophy of his French colleagues—was situated firmly within the British tradition of the natural sciences generally and astronomy particularly, supporting a broad interpretation of Christianity.[8]

William and his family soon concluded their Parisian visit and returned to England, where the astronomer resumed his telescopic surveys of the heavens, eventually making the discoveries that Laplace would take as confirmation for the very theory that he had argued with Napoleon. Within two years of the Herschels' return, the first consul had crowned himself emperor, Britain and the new French Empire were again at war, and England itself was threatened with invasion. The military and political conflict that had dominated the national context throughout John Herschel's childhood would continue to his time at university. Laplace, however, would survive Napoleon's ultimate fall and retain his influence throughout the restoration of the Bourbon monarchy. Though William never returned to France, John would resume conversations on the structure of the universe with the elder statesman of French science when he returned as a young man.

At ten years old, John Herschel no doubt had a very different perspective on his time in Paris than his father, but William's experiences there colored his own. The organization of science, the overarching scientific program of Laplace and his colleagues, the professional meetings of the Institut, the bourgeoning geological debates—all these showed a pursuit of knowledge different from what John would experience in Britain. In addition, the trip allowed William to see firsthand the role that the new French mathematics played in making the scientific accomplishments of Laplace and his school possible, and he determined that whether or not such tools were widely available in England, they would become a feature of John Herschel's education. The mathematics of Laplace's France, nurtured under the autocracy of Napoleon, would ultimately be the first step in the younger Herschel's reform of British science.

CHAPTER 1

A CAMBRIDGE DISCONTENT

Most of the advantages which science has received, are to be attributed to improvements made in the language of Analysis.

—ROBERT WOODHOUSE,
PRINCIPLES OF ANALYTICAL CALCULATION, 1803

I think a Cambridge education has for its object to make good members of society—not to extend science and make profound mathematicians.

—SIR FREDERICK POLLOCK, SENIOR WRANGLER, 1806

Charles Babbage was a difficult friend at the best of times, but at the end of his student years John Herschel's patience with him had run out. It was not the first time Herschel had been exasperated by Babbage's impetuousness, and it certainly would not be the last. Babbage, the son of a wealthy banker, was a student at Trinity College and a year behind Herschel, who was at Trinity's rival and neighboring college of St John's. Both Herschel and Babbage had arrived at Cambridge to find the university, to their disappointment, backward, conservative, and at times downright infuriating in terms of pedagogy and curriculum. The two friends agreed on what needed to be done to improve this situation, but their methods for achieving those improvements differed widely. Herschel, the only son of the world's most famous astronomer, proved a conservative reformer. Conscious of his family's marginal social status, he learned the mathematical exercises required for the all-important culminating Senate House exams, despite their tediousness and outdatedness. His dutiful labors were rewarded, as everyone expected, by his achieving the position of highest scoring, known as the senior wrangler.

Babbage, by contrast, had no patience for the slow reform of systems he saw as archaic and counterproductive. Despite being as skilled in mathematics as Herschel, Babbage refused to play by the university's rules. One of those rules was that prior to sitting for the Senate House exams, students complete an act of disputation, a formal oral exam that was by this point a vestigial tradition. Babbage chose to turn his act into a display of revolutionary pique. Instead of choosing an innocuous topic to defend, as Herschel had done by selecting a passage from Newton's *Principia*, Babbage shocked his examiners by addressing the radical topic of whether God was a material agent, in essence arguing against God's supernatural nature. His opening remarks were met with the angry cry of "Descendo," or *fail*, from the head examiner and cost Babbage the chance to perform in the Senate House exams.

When the normally unflappable Herschel found out about Babbage's performance, the two young scholars had a heated argument. Herschel, cornering Babbage in a friend's room, berated him for his behavior. They were both frustrated with the system, Herschel admitted. They were both disappointed with what Cambridge had to offer as far as mathematical and scientific training. Their instruction lagged behind important developments taking place outside Britain. But for now, Herschel insisted, they needed to play the academic game. Babbage's performance not only embarrassed himself; it endangered the mathematical revolution he and Herschel had been cultivating at Cambridge.

Babbage was unrepentant. Despite their close friendship, Babbage never succeeded in fully following Herschel's appeals for more circumspection or patience. "Herschel," Babbage said that evening as he stormed out of the room, "you are as great an Ass as myself." As Babbage slammed the door behind him, Herschel feared he was watching both his friendship and his aspirations for reforming British natural philosophy falling down around him almost as soon as they had begun.[1]

AN ASTRONOMICAL LEGACY

When John Frederick William Herschel arrived as a student at Cambridge University, before he met Babbage and four years before their stormy exchange, he did not intend to change the face of the scientific endeavor at the university and throughout the world. Almost all students who entered Cambridge (who were only men until the first women students—including Herschel's daughter Constance—were admitted in 1869) came not for academic rigor but rather for enculturation into privileged society and to establish professional connections that would hope-fully lead to careers in the law, the church, or medicine. Despite having been the home of Isaac Newton a century before, Cambridge had settled into a sleepy indolence, satisfied with preserving Newton's tradition in mathematics and providing a veneer of education in classical literature, natural philosophy, and theology.[2]

Herschel already stood out from the other sixteen- and seventeen-year-olds arriving as first-year students. By the time John Herschel was born on 7 March 1792, his father William's discoveries and instruments had given William a place among the world's scientific elite, but a unique one. Many simply did not know what to make of this musician-cum-astronomer and his unusual telescopes. In particular, his claims for their almost unbelievable magnifying powers were so outrageous to some that they made him "fit for Bedlam." It also did not help that William was not hesitant to share his more unusual astronomical beliefs. He was sure that the universe was filled with life, which was not so unusual for the time, but William thought it was possible to observe evidence of this life on the moon and believed even the surface of the sun itself was inhab-ited.[3] Others recognized him as an excellent observer and instrument maker but not necessarily a scientific natural philosopher. As one satiri-cal writer put it just a few years before John's birth,

> When on the moon he first began to peep
> The wond'ring world pronounced the gazer, deep
> But, wiser now th' unwond'ring world, alas!
> Gives all poor Herschel's glory to his glass.[4]

This distinction, of being a talented artisan rather than a true mathemat-ical philosopher, haunted William and had important implications for his son's education.

In retrospect, William's greatest accomplishment was, along with his sister Caroline, changing the focus of astronomical research. Tradi-tionally, astronomy had been the study of the positions and motion of objects in the solar system. This included the path of the moon against the background stars, useful for calculating longitude at sea, and the position of planets and comets in the sky, useful for testing the limits of the law of gravity as formulated by Newton. Things beyond the solar system such as the curious smears of light glowing like luminous clouds in William's telescopes and questions about the nature of the stars them-selves were outside the realm of proper, measurable astronomy. Together, Caroline and her brother created catalogs of those faint, gauzy clouds called nebulae and the hundreds of double stars William detected in his nightly sweeps of the sky, objects that no one else had instruments to see. Caroline and William became pioneers of a new sidereal astronomy that reached beyond the solar system.

To William this sidereal universe was anything but static. He wanted to understand the processes and connections among stars and nebulae the same way naturalists understood relationships between animals, fossils, and the layers of Earth. The cosmos was a dynamic, living system. His work, he explained, seemed to throw the universe "into a new kind of light." The heavens were "seen to resemble a luxuriant garden." Just as a careful observer walking through a forest or garden could deduce the life cycles of various plants by observing them in their different stages of growth, so William believed he could understand the life cycle of stars and nebulae, even though progression from one stage to the next might take entire ages of the world. Yet he lacked the mathematical tools to make his vision a rigorous scientific theory.[5]

William was fifty-three years old when John was born. The elder Herschel had put off marrying, consumed first by his musical career and then by his astronomical pursuits. For decades, he and Caroline lived and worked together, with their brother Alexander coming frequently for extended stays. When William became the king's personal astronomer, the family moved to new lodgings in a village called Slough, close enough to Windsor Castle that the royal family could call him to court whenever they wanted an evening of astronomical diversion. Slough was growing from a sleepy village to an important stop on the stagecoach line from London and the resort town of Bath, and William set up shop in a dilapidated home that would eventually become known as Observatory House.

In Slough a local family called the Baldwins had built a small fortune running an inn as well as the first daily coach service between London and Bath. William's new landlord was a Baldwin with a widowed daughter named Mary Pitt, who lived nearby. Mary and the distracted astronomer began a slow courtship, which culminated in marriage six years after William's arrival in Slough. Caroline, who had recorded hundreds of nights of her brother's observations, processed the raw data into calculated results, organized his catalogs of stellar objects, and at times even drafted his scientific papers, was pushed out of Observatory House to find lodgings in a series of nearby cottages. Though no longer part of the immediate household, Caroline would be an important part of John's childhood.[6]

John was raised and educated under the watchful eye of his parents, who were both advanced in age for the period by the time he was born, as well as the adoring gaze of his aunt Caroline, who wrote in her journals with obvious pride of John's antics climbing the scaffolding of the giant forty-foot telescope and ruining carpets with his chemistry experiments.[7] His childhood was filled with astronomical discussion, the polishing of great mirrors, and the occasional evenings his father would pass in all-night observation. Though William's period of active observing was largely passed by the time John came along, many travelers stopped at Slough hoping to meet the famous astronomer and spend an evening stargazing. Throughout John's childhood, the guestbook of Observatory House included members of the royal family, who often called on their personal astronomer, as well as philosophers, novelists, and mathematicians. John's uncle Alexander, who came for several weeks each year to assist his brother, was another fixture. The frameworks John's father had raised up were all around him, from the literal wooden scaffoldings of the large twenty-foot reflecting telescope and the enormous forty-foot one, to the figurative frameworks of those who came to compare ideas with William on the nature of the universe.

John grew up in a world where his family name had been synonymous with astronomical discovery since before he was born. This legacy followed him when as a sixteen-year-old he visited Cambridge for the first time. That scientific legacy, however, also included an awareness of the deficiencies in his father's education that despite his fame kept William's work somewhat peripheral to the scientific community. At Cambridge John would have opportunities for the mathematical training and social advancement his father, who had arrived in England as a refugee, never had. But before Cambridge, years of preparatory education were necessary.

PREPARING FOR CAMBRIDGE

It was not so much his ideas about extraterrestrial life or claims of the magnifying powers of his immense instruments that kept William Herschel's work outside the astronomical mainstream during John's childhood. It was rather William's lack of extensive mathematical training, which meant that his most important discoveries never had the mathematical rigor that was coming during this period to define scientific theory. William's mathematics was self-taught, snatched from books he purchased and pored over during his days as a musician in Bath. Though he read and grappled with the fluxional calculus, the apex of British mathematics developed by Newton the century before, this method was quickly being superseded by advancements outside England, and William knew it.

Fluxions were Newton's means of formalizing the mathematics he created for his great gravitational synthesis of the *Principia*. To quantify concepts like the orbits of planets and comets in the solar system, Newton needed to treat physical properties (like velocity or direction) that varied over time. The rate of change of any such property at a specific moment or location was known as a fluxion. (Today we call the analogous concept a function's derivative and the field of mathematics that deals with them calculus.) The concept of fluxions was based on physical or geometrical ideas: bits of time, distance, or other quantities becoming infinitely small. This, however, was becoming a problem for many mathematicians during William's lifetime. For one thing, some mathematicians felt it was not logically meaningful to talk about physical quantities becoming infinitely small. There was a desire to create systems of mathematics that could do the work of fluxions but remain "untainted" by vague concepts like infinitely small quantities. One way to do this was to create a calculus completely independent of geometry and based solely on algebra.[8]

As William Herschel was making his early astronomical discoveries, before John was even born, Newton's fluxional calculus was quickly being replaced by developments elsewhere in Europe and particularly in France. Britain was disconnected from these mathematical developments, a fact exacerbated by the backdrop of war with France. The two countries were at war from 1783 to 1802 and again from 1803, when an invasion of Britain seemed immanent, until Napoleon's final defeat at Waterloo in 1815. To put the scale of these conflicts into perspective, the wars with France were more than twice as long in duration as World Wars I and II together and, as far as Britain's involvement, nearly as extensive geographically.[9] This series of wars meant not only were mathematical and scientific developments in France hard to keep abreast of in England, especially as the British blockade of the Continent made books difficult to transport, but these developments were also viewed with suspicion. Why would English patriots be interested in importing French learning with potential revolutionary overtones? Most mathematicians in England remained firmly committed to the legacy of Newton and saw any challenge to fluxions as a betrayal of this legacy.

Meanwhile, as William Herschel had seen on his visit to Paris, mathematicians abroad and especially in France were developing powerful new methods to address and explain physical phenomena. Unlike traditional methods (such as fluxions) based on geometrical notions, Continental analysis moved away from physical representation. An exemplar of this approach was Lagrange's *Mécanique analytique* (published in two volumes in 1788 and 1789), a treatment of physics whose preface boasted it did not contain a single diagram. In analysis, geometric representations of things like motion or velocity were unnecessary. Analysis treated mathematics as a grammar with logical rules instead of a picture that represented the world.

This new approach was incredibly fruitful. Teachers like Sylvestre François Lacroix wrote textbooks training the next generation of French public servants and engineers in the methods of analysis. Calculus was re-founded on a purely abstract foundation, without reference to motion or velocity. The culminating triumph of the analytical approach was Laplace's *Méchanique celeste*, which offered a view of the entire universe, from the motion of the moon to the long-term stability of the solar system, through the lens of analysis. All astronomy up to that point, including many of William Herschel's discoveries, was presented in a purely analytical form. In France, mathematical analysis was triumphant.[10]

The mathematical situation in Britain was woeful by comparison. As Thomas Young (1773–1829), one of the leading natural philosophers in England during this time, grieved, "We do not believe that ten persons in the universe have read Laplace's *Mécanique céleste* as it ought to be read."[11] Hardly any of those hypothetical ten were in England, as few philosophers had the training necessary to make sense of the text. Even if someone in England wanted to study the work, the war with France made it difficult to get it or other mathematical texts. Those that did make it to England were hard to find, incomplete, and expensive. In France, mathematical analysis was transforming the way people saw the universe, but this transformation was leaving Britain behind, and the largest telescopes in the world were not enough to keep pace.

This all made William keenly aware of the need for careful mathematical training in John's education. There was no one to turn to in England for such instruction, but natural philosophers in Scotland maintained closer connections with French developments. William already had links with Scottish natural philosophers such as David Brewster (1781–1868) and John Playfair, who he had met on various trips to Scotland. Scotland remained more open to French influence than England, and William, who saw himself as a member of a scientific community that transcended national boundaries, drew on his Scottish connections for the education of his son. As a result, John would end up drastically overprepared for mathematics at Cambridge.

Though William deferred to his wife, Mary, in directing the classical schooling expected for a son of their social standing (and Mary's inherited fortune made elite private education possible), he took an active role in the mathematical side of that education. At the age of five, John was sent to board with a Mr. Atkins and soon after was sent thirty miles down the road to a boarding school in Newbury. At the age of eight, John started at Eton College, not a college at all but rather a famous boarding school that prepared sons of the aristocracy and upper classes for Cambridge or Oxford. William insisted that mathematical training be a major part of John's education. This was not the case at Eton and may be one reason John spent only three weeks there (though Herschel family lore also tells the story of Mary visiting and witnessing John about to fight an older boy). After Eton, John was transferred to a boarding school in Hitcham House, five miles west of Slough, to be tutored in the classics by Reverend George Gretton (1754–1820), formerly a fellow of Trinity College, Cambridge. William had private discussions with Gretton about the mathematical curriculum at Hitcham House, which also proved insufficient for William, so the Herschels employed at least two in-home tutors to provide John with additional mathematical training.

It was the second of these tutors, a Scottish scholar named Alexander Rogers, who had the most influence on John's early mathematical education. Rogers was likely recommended by William's friend Patrick Wilson (1743–1811). Wilson, professor of astronomy at Glasgow University until 1799, had corresponded with the elder Herschel for years as an informal math tutor. William had met Wilson and other Scottish natural philosophers when he first traveled to Scotland in 1783 to be awarded an honorary degree at Edinburgh University and again in 1792 for an honorary degree at Glasgow University, leaving a three-month-old John at home. When John was fourteen, the Herschels paid a substantial sum for Rogers to travel from Edinburgh to lodge at Slough and work with John outside of John's normal school days. Besides mathematics, Rogers also tutored John in French, a skill necessary to read the latest mathematical publications.[12]

John was particularly interested in the ne plus ultra of the analytical approach, Laplace's *Mécanique céleste*. If John's father had peered farther than anyone into space, it was Laplace in the *Mécanique céleste* who had gone further than anyone in mathematically perfecting astronomy. For John, who no doubt recalled his father's accounts of the hours spent in conversation with Laplace at the Institut or his ornate home in Paris, Laplace represented the culmination of mathematical astronomy. Rogers himself admitted to being able to make it through only the first portion of Laplace's work and urged John to start instead with Lagrange's *Mécanique analytique*, which was simpler and would be good preparation for Laplace. To understand the universe, John realized, one had to first understand French mathematics. Cambridge, he hoped, would be a place to continue to develop and expand these mathematical skills.

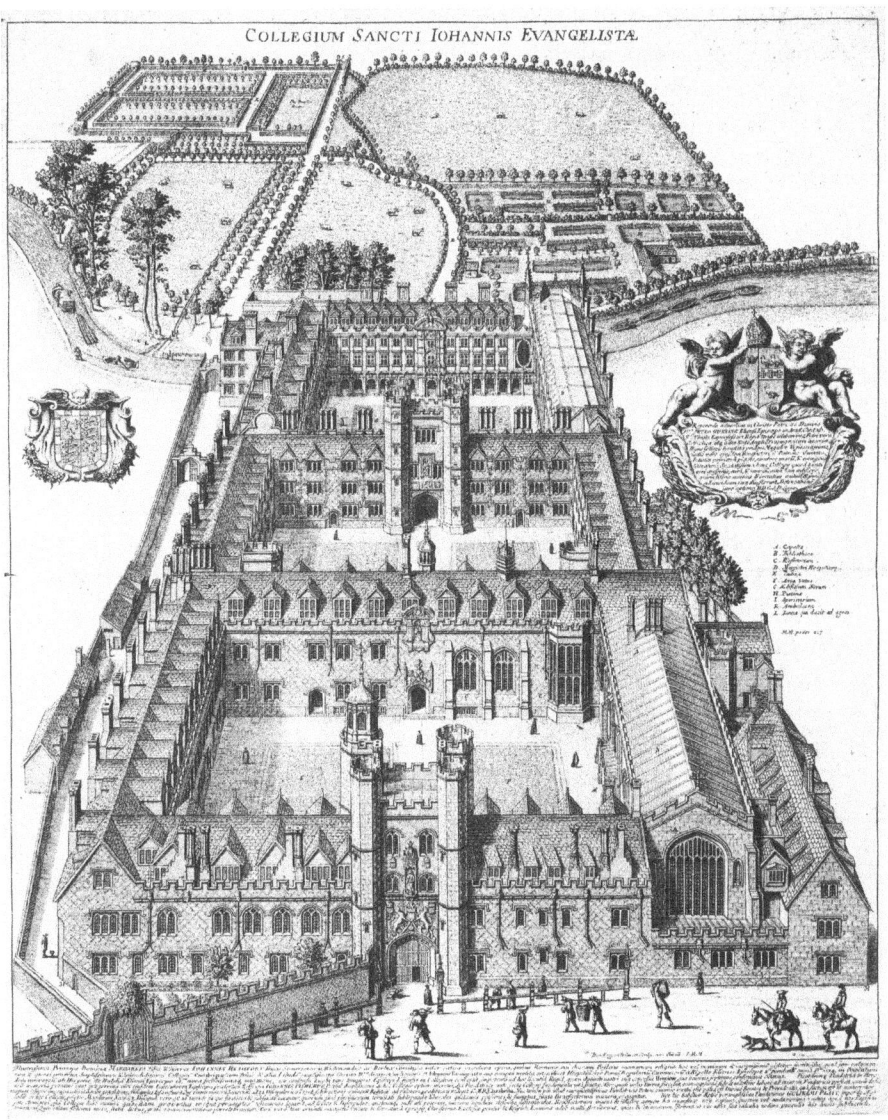

FIGURE 1.1: Engraving of St John's College, David Loggan, c. 1685. Public domain.

COMING TO CAMBRIDGE

When John Herschel matriculated at St John's College, Cambridge, at the start of Michaelmas term in October 1809, he was seventeen years old and had been exposed to the mathematical work of Lacroix, Lagrange, and Laplace—none of whom were taught at Cambridge. Though it lacked the size and prestige of Newton's Trinity College, the Herschels had settled on St John's College as the best fit for John. It was among the most powerful in the university, in no small part because of its enormous wealth, and was growing rapidly. Undergraduate admission to St John's had been about forty students in 1800 and would more than double to over ninety by 1820.[13]

St John's was known for the role that its alumni, especially the politician William Wilberforce (1759–1833) and the campaigner Thomas Clarkson (1760–1846), had recently played in passing the 1807 Act of Abolition outlawing the British slave trade. The college was where the British abolitionist movement had first taken root, making Cambridge an antislavery stronghold, a fact that resonated with Mary Herschel as an ardent abolitionist. And though William Herschel had never attended university himself, his earliest and best supporters in scientific society, the Bath physician William Watson (1744–1824) and Astronomer Royal Nevil Maskelyne (1732–1811), were both Cambridge men. St John's would prove an ideal fit for the young Herschel.[14]

As the son of a non-aristocratic, well-to-do family, Herschel entered the university as a "scholar," meaning his tuition fees were reduced by half, coming to about thirteen pounds a term. His parents provided him an allowance of sixty pounds each quarter, so there was plenty left over after expenses for trips to the theater, flute lessons, purchasing books and equipment, and even learning to ride the velocipede, an early form of bicycle, on some of the city's first paved streets. But the seven-hour journey from Slough to Cambridge took Herschel out of the shadow of the telescope in a physical sense alone: as soon as anyone, from fellow students to professors, heard his name, they knew exactly who he was. Herschel could not escape the expectations that came from being son of the world's most famous astronomer.[15]

Herschel hoped to continue the mathematical training he had enjoyed under Rogers's tutelage, but Cambridge in the first decades of the new century existed not for research or discovery but for vocational training. Studied texts were selected not so much for intellectual rigor or application in advancing natural philosophy as for their perceived ability to hone the intellect, sharpen clarity of thought, and instill a sense of the orderliness of nature and its creator through the study of natural theology. John would pass his time reading Hobbes, Locke, and Hume, as well as Scottish philosophers like Thomas Reid (1710–1796) and Dugald Stewart (1753–1828), attending chapel, and sitting for the occasional lecture. Ironically, however, ultimate success in this academic environment was nonetheless based on a culminating series of mathematical tests: the Senate House exams, named for the building in which they were taken. The Tripos (as the exams were more familiarly known, possibly because of the three-legged stools the students once used) had become a three-day ordeal undertaken at the end of a student's tenth term, with a fourth day of testing added the year before Herschel's arrival. (There were three terms each year, which meant the Tripos were usually taken a third of the way into a student's final year at Cambridge.) Two weeks after the Tripos followed an additional, optional exam known as the Smith's Prize, in which students jockeyed for the fellowships that would allow them to remain at Cambridge after completing their undergraduate studies. In theory and often in practice, this exam system made Cambridge a meritocracy, allowing students of even modest backgrounds who placed top in the exams to be eligible for lucrative church and university positions.[16]

Three years of preparation at Cambridge for the culminating mathematical exams might have appealed to someone like Herschel, who was already reading the latest mathematical work from the Continent, except that these exams covered a completely different and outdated form of mathematics. There would be nothing on the exams, Herschel quickly realized, mathematically or scientifically relevant or engaging. It quickly became apparent that Herschel had already mastered the level of mathematical expertise needed to succeed on the tests before he had even arrived at the university. In this sense, Herschel's preparations with Rogers had set him up for disappointment with mathematics as he found it at Cambridge. Instead of elegant algebraic abstractions, Herschel found elementary geometric figures. Instead of equations, there were traditional and unwieldy ratios and proportions. Instead of clear mathematical generalization, he found tedious fundamentals. Herschel complained to his former tutor about the "irksomeness" of his studies, which required him to retread ground he had already covered. Far from preparing him for new investigations or challenging and expanding on what he had already learned, Herschel was expected to devote his time to preparing for success in a test dominated by outmoded mathematics.

There had been some attempt at mathematical reform before Herschel's arrival. The Cambridge fellow and mathematician Robert Woodhouse (1773–1827), for instance, wrote a text in 1803 on the foundations of calculus using analytical methods and published two papers on the topic in the *Philosophical Transactions of the Royal Society*, Britain's premier scientific publication. In his text Woodhouse tried to base calculus entirely on algebraic principles. He acknowledged that geometrical figures were useful for clarity but emphasized his preference for analytical expression. In order to generate useful formulas and move beyond simple demonstration, Woodhouse explained that "the geometrical method, at some point or other, must be abandoned, and that of algebraic signs adopted." In these papers Woodhouse, who had dominated the Tripos as a student in 1795, argued against those who limited mathematics to concepts with physical meaning, a constraint that ruled out ideas like negative and imaginary numbers.[17] Despite his views, though, Woodhouse did little to introduce analysis into the Cambridge curriculum. His work had negligible influence on formal instruction at the university; nonetheless, Herschel found it and was encouraged to continue this mathematical approach. Herschel's very first (unpublished) paper applied Woodhouse's approach to spherical trigonometry, and his first published papers drew directly on Woodhouse's work.

During Herschel's Cambridge years, the continuing conflict with France cast its shadow on mathematical pursuits. As Britain blockaded the Continent to choke off Napoleon's forces, it became prohibitively expensive to keep up with mathematical work published in France. During the height of the Napoleonic wars, a student looking for Lacroix's standard textbook on calculus found it could only be purchased for the exorbitant cost of seven guineas. (During this period, a guinea was worth about one pound sterling, meaning the textbook in current dollars would have cost more than $600.) The student was Babbage, who, after a slight hesitation, bought it.[18]

But this shadow was more than purely economic. Why would any good Briton be so interested in the mathematics that had trained a generation of French artillery officers and engineers in the army now pummeling Britain and its allies in the field? Not long before, in the years following the French Revolution, antirevolutionary sentiment in England had run high. It had been quite dangerous to hold suspected sympathy to a regime that had disposed and executed a royal family. Joseph Priestly (1733–1804), for instance, a well-known chemist and natural philosopher, had his house ransacked in 1791 by an angry mob in Birmingham during the height of antirevolutionary riots. Being perceived as too interested in French ideas could have explosive repercussions.

By Herschel's Cambridge days, this antirevolutionary fervor had calmed somewhat, largely because, with the rise of Napoleon and the jettison of France's more liberal revolutionary ideals, Britons became more unified in their opposition to France. There was less chance that anyone would espouse pro-French sentiment, in any field. Herschel, however, eschewed these nationalist ideals. He had been raised to see natural philosophy, and in particular the mathematics that aided it, as transcending national boundaries and conflicts. The disruption of the wars with France to the community of natural philosophers and mathematicians would color Herschel's perspectives on scientific collaboration for the rest of his life. Yet in a nation largely unified against all things

French, those sympathetic to French mathematics and eager to spread it at the expense of Newton's mathematical legacy appeared dangerously radical.

In spite of this, the educational seeds William had made sure were planted in his son's early education began to bear fruit, quickened by the younger Herschel's frustrations with the state of mathematics in the nation's foremost university. Traditional mathematics taught traditionally was the core of the liberal education that formed Cambridge's very raison d'être.[19] In Herschel's mind, though, Cambridge in particular and Britain more generally needed analysis. As radical as French mathematics might appear, the time was ripe for a revolution that would challenge the Cambridge educational system and set mathematics on a new trajectory. Yet the catalyst that set things in motion came not from Herschel but rather from Babbage. And, as was often the case with Babbage, it began as a joke taken too far.

THE ANALYTICAL SOCIETY

Despite frustration with his formal mathematical instruction, Herschel continued his own studies on top of the prescribed Cambridge curriculum. Within his first year at Cambridge, his former tutor Rogers admitted, "Your mathematical knowledge already surpasses mine." Rogers encouraged his former pupil to expand his knowledge, telling him mathematics would be "an inexhaustible fund of rational improvement" as it ordered Herschel's thinking. This was the standard Cambridge line on mathematics as central to the liberal arts. On the other hand, Rogers emphasized mathematics as "a means of extending the boundaries of science," a tool for discovery. This tension between the roles of mathematics as a system to shape thought or a tool to investigate nature was one Herschel wrestled with throughout his Cambridge years and beyond.[20]

Part of the problem when it came to pursuing advanced mathematics was that at this point in Britain there was no such thing as a professional mathematician or scientist. The word *scientist* had not yet even been invented. (Herschel's friend and fellow Cambridge graduate William Whewell coined the term decades later.) Beside positions as college fellows or professors, there were no vocations in which pursuing natural philosophy or mathematics could be one's full-time livelihood. One possibility was to get a job as a mathematics teacher in the few private academies that offered advanced training in these topics, but Rogers, trying to make ends meet in Edinburgh, warned Herschel away from this path: "If a person has acquired a taste for any branch of science, . . . the readiest way to become disgusted with its pursuit is to undertake the situation of assistant in an academy where it is taught."[21]

Most often those who pursued mathematics or natural philosophy did so from positions within the church, usually as curates or vicars of small parishes. As Herschel's father would argue when it later became a point of contention between them, a church living provided opportunity to give moral guidance (regardless of how one felt about church doctrines) along with the freedom and leisure to pursue science. Indeed, John Herschel's Cambridge education was the ideal foundation for such a life. For centuries, Oxford and Cambridge had functioned more or less as seminaries, and for many students a bachelor of arts was the first step toward ordination. Fellowships in the colleges still held vestiges of ecclesiastical positions, with most fellows holding church postings and all being required to remain unmarried. The generous church livelihoods eventually secured by many friends and colleagues from Herschel's student years almost always came about through their Cambridge connections. St John's in particular had a tradition of preparing clergymen: during the 1700s they provided more clergy for the Church of England than any other Cambridge college. This seems to have been Herschel's assumed path as well. A close friend consistently referred to him as a "Devil Fighter" and "nascent apostle" but warned that Herschel would never rise far in the church because his sermons would be too bland to frighten any followers of the Devil.[22]

Regardless of plans for the future, at Cambridge Herschel continued to pursue mathematics far beyond the required level. His first publications appeared the spring of his third year as a student. These were two short mathematical notices in the *Journal of Natural History, Chemistry, and the Arts*, a periodical published by the London-based editor William Nicholson (1753–1815). Nicholson's journal printed an eclectic mix of news, reviews, and correspondence from readers alongside copies and excerpts of published scientific papers. Contributors to the journal might be completely unknown to the scientific community but nonetheless could expect to have their articles published quickly and garner response. This contrasted with a recognized publication like the Royal Society's *Philosophical Transactions*, which printed no letters, no scientific news, and no notices of recent books. Authors could not even submit to it directly; the papers it included could only be printed after first having been presented at a society meeting.[23] For a young scholar, Nicholson's journal offered a chance to share mathematical work with a wider audience.

Herschel's first published work, in an 1812 issue of Nicholson's journal, was published under the pseudonym "A Lover of the Modern Analysis." Titled "Analytical Formula for the Tangent, Cotangent, Etc.," it represented the work of a disciple of Woodhouse. In the short four-page paper, Herschel showed how to produce trigonometric expressions by algebraic methods, giving the derivation of functions expressed as sums of series. His second article appeared in the next monthly issue. This time "A Lover of Modern Analysis" had become "Analyticus." In this second, brief article Herschel derived algebraic series related to expressions for pi and explicitly referenced Woodhouse's *Trigonometry*. Both papers were analysis in action: pages of purely algebraic expressions with no diagrams, demonstrating the power of analysis to derive and express functions previously understood in geometric terms.[24]

So far though, Analyticus was working alone. But on his return to Cambridge for the second term of his third year, something changed. For the first time, Herschel found himself among others with similar mathematical interests and passions. A handful of students came together late that spring for meetings of a club that started out as a joke but became a defining facet of Herschel's life and of the history of mathematics in Britain.

It started with a campus controversy. In early 1812 a debate raged over whether the university would support the founding of a chapter of the interdenominational British and Foreign Bible Society. Most who cared agreed that the Bible should be distributed, both to the poorer classes in England and to British subjects abroad. The problem was that the society, outside the control of the Anglican establishment yet far more successful than the official Church Missionary Society, planned to distribute scripture without any sanctioned commentary or the Church of England's Book of Prayer. The subsequent controversy became over whether the Bible should be distributed alone and left to readers to interpret themselves. Campus authorities such as James Wood (1760–1839), the famously conservative master of St John's, opposed the two hundred students who met to form a Cambridge branch of the society and distribute Bibles without an accompanying interpretive framework.[25]

Chances are that Herschel's friend Babbage had no strong opinions one way or another on the question of distributing scriptures with or without commentary. However, when a fellow student of Trinity College, Michael Slegg (1791–1889), commented on the debate to Babbage in early May of 1812, it gave Babbage the idea for a fantastic spoof. Babbage announced the formation of a society to distribute a different holy text, Lacroix's 1802 *Traité élémentaire du calcul différentiel et du calcul intégral*, a textbook central to teaching mathematical analysis and nearly impossible to find in Britain. Babbage's joke turned the controversy on its head: when it came to mathematics, Cambridge was no longer the enlightened city distributing the Word of God to the unenlightened at home and abroad. Instead, Cambridge was the place that needed enlightenment, and a society would be formed to bring the gospel of analysis from Continental Europe, where it thrived, to the British back-water where knowledge languished. Babbage quipped that such a society would transcend the "dot-age" of the university and instead preach the "d-ism" of the Continent. (In the notation of fluxions, a dot was placed over functions to indicate their derivative, whereas derivatives in the Continental notation were indicated by placing a d in front of the appropriate variables, the standard notation today.)

Slegg passed Babbage's joke to another mutual friend, Edward Bromhead (1789–1855). Bromhead was aristocracy, son of a baronet and baroness, born in Dublin and educated at the University of Glasgow before coming to Cambridge. He had recently graduated but remained at the university for a time before leaving to study law in London. Bromhead, himself a skilled mathematician, loved the idea and wanted to take it further, offering use of his rooms at Caius College for a meeting place. Once the ball was rolling, the society quickly evolved beyond the joke. The first meeting mustered eight students. Original members of what would become known as the Analytical Society included George Peacock (1791–1858), who went on to become an important mathematician and eventually dean of Ely, a powerful church position associated with Cambridge; Richard Gwatkin (1791–1870), ultimately senior wrangler of his class and like Peacock going on to a career in the church; and John Whittaker (1790–1854), eventually one of the founders along with Herschel of another transformative society, the Astronomical Society. Herschel, who by this time knew Babbage, was immediately brought into the group's orbit, and Herschel and Babbage along with Bromhead quickly became the driving force of the new society. They agreed to meet the first Monday of every month to discuss mathematical topics.[26]

Because there was little opportunity for authorized college activities outside of lectures and chapel, informal student societies were common at Cambridge during this time. Varsity sports beyond cricket and rowing did not yet exist, and the Cambridge Philosophical Society (which caused such a controversy that for a time it was forcibly suppressed) was still years off. Babbage had personal experience founding societies, having already founded a Ghost Society and a society devoted to skipping chapel to go boating. He and Herschel also met informally with other friends for philosophical chats over breakfast, but none of these informal groups left the same impact as the Analytical Society.

For Herschel, the effect of the Analytical Society was transformative. Babbage, Bromhead, and the others provided fellowship and encouragement for mathematical work. The society filled a gap in the collective Cambridge experience of these young mathematicians. Despite the evangelizing context of its formation, the society's purpose was not originally to reform mathematics teaching at Cambridge. That would come later. Initially, it simply provided a sympathetic audience and an excuse to pursue analysis for its own sake. Ironically though, the only members who could actively participate were those with enough mathematical knowledge to be already well prepared for the exams and who could spare time and energy for mathematics that had nothing to do with their formal education.

Herschel had skill, time, and energy to spare. Once the Analytical Society was up and running, even breaks from Cambridge did not slow him down. He was no longer "Devil Fighter" in correspondence with friends. Instead, it was "Dear Mathematician," though a friend writing from Scotland wondered whether Wood, the master of St John's, had yet "succeeded in blowing the Analytical Society to the Devil." For Herschel, the society quickly became a model for collaborative work fueling his research, an approach he would cultivate for the rest of his career. His letters from the summer after the society began meeting were filled with mathematical correspondence. "I rejoice," he wrote to Babbage in a letter recounting and comparing mathematical results, "to find that you are still labouring in the cause of reason and truth."[27]

Rogers had emphasized mathematics to his former student as a peaceful diversion that provided glimpses of physical truth. But Herschel's work took him further, to embrace analysis as the *basis* of truth. It was not simply abstract mental exercise. The new mathematical tools of analysis were the keys to understanding the relationship of reason to the natural world and ultimately the functioning of human reason itself. Herschel insisted the society needed to grow beyond the university and endure after their graduation. It was too important to remain simply a *Cambridge* Analytical Society, a diversion for a group of eccentric and like-minded students. The Analytical Society needed to expand to transform mathematical practice throughout Britain. In this, despite his best efforts, Herschel would ultimately be disappointed.

The summer after the formation of the Analytical Society, Herschel began work on his first major publication. He had contributed papers for meetings of the society in the previous term, but this would be his first formal paper, a transition from the informal meetings of the society and the eclectic pages of Nicholson's journal to the mouthpiece of Britain's oldest scientific society. The Royal Society of London had been founded by King Charles II in 1660 and was the birthplace of the practice of gentlemanly science and experimentation, where early luminaries like Robert Boyle (1627–1691), Robert Hooke (1635–1703), Edmond Halley (1656–1742), and Isaac Newton shared their experimental results and new ideas about the universe. The pages of the *Philosophical Transactions*, published since 1665, were the means by which scientific practitioners and natural philosophers kept one another abreast of their latest discoveries and how an educated elite with leisure and interest stayed connected to developments in science.

William Herschel had risen to fame by sharing his discovery of his new planet with the Royal Society's president, Sir Joseph Banks (1743–1820), and William subsequently published his most important papers in the society's *Philosophical Transactions*. The journal carried only work that had been submitted or sponsored by fellows and presented in meetings and treated everything from new mathematical formulas to astronomical observations to chemistry and geology. John Herschel's first submission to the journal would be a scientific rite of passage, a way of presenting himself, through his mathematical work, as deserving a place among Britain's scholarly elite.

Home in Slough and again in the shadow of his father's great telescope, Herschel composed a paper on analytical mathematics linked to questions of planetary orbits. Titled "On a Remarkable Application of Cotes's Theorem," the paper began without explanation or preamble and launched directly into mathematical explication. Herschel showed that Cotes's theorem, a geometrical relationship derived by the English mathematician Roger Cotes (1682–1716), had important implications for solving equations related to conic sections. The paper is a thicket of mathematical expression (raising the question of how it was ever read in a Royal Society meeting), and Herschel provided no explanation of what its physical application might entail besides the barest nod toward astronomy. Readers would have recognized that conic sections, representing the paths traced out by bodies moving according to Newtonian gravity, provided a link between the astronomy of the father and the mathematical passion of the son. It was fitting then that William communicated the paper to the society on his son's behalf (since John was not yet a member) at a meeting of 12 November 1812.

The paper shows the extent of Herschel's new way of viewing mathematics. Even when treating equations that represent the motion of physical objects, geometrical representation is no longer useful. Theorems derived algebraically, Herschel explained in the conclusion of the paper, "become for the most part complicated and unintelligible when geometrically enunciated." Analysis makes it easier to work with relationships that would be too complex geometrically. Despite physical applications, Herschel took pains to make it clear he was not doing geometry. The importance of his results was not in the physical or geometrical curves that his equations represented. Instead, the properties of the equations he derived, he explained, were "properties rather of the equations of the conic sections, than of the curves themselves." Truth was in the analysis, in the relationships between numbers and functions, not its connection to the physical world.[28]

This distinction between mathematics and the physical world is even clearer in Herschel's comments in the paper regarding mathematical notation, which hinted at the radical program the Analytical Society would later espouse publicly. Just as a spoken language encodes assumptions and can even make certain modes of thinking more possible or more difficult, mathematical notation can do the same. When Herschel introduced the inverse cosine function in his paper, for example, he took pains to divorce it from its traditional geometrical trappings. Sine, cosine, and tangent (and their inverse functions) rather than being geometrical concepts needed to be understood as "merely *characteristic marks* to signify certain algebraic operations performed." For Herschel, even the most familiar trigonometric functions were more truly defined as algebraic operations, not geometrical representation.[29]

Herschel's first Royal Society paper was a model for the kind of gentle revolution he was beginning to pursue in mathematics. Unlike Babbage, Herschel felt the importance of working to reform mathematics *within* existing scientific and educational systems. In his paper, he hinted at some of the radical views of the Analytical Society, but he chose to apply analysis to the work of Cotes, a figure safely in the Newtonian tradition and, like Newton himself, a Cambridge mathematical hero. With a paper on Cotes's theorem, Herschel placed himself in a recognized mathematical tradition even as he applied Continental methods and offered new ideas regarding the nature of mathematics itself.

Herschel's paper was completed the day he left Slough at summer's end to return to Cambridge for his final year as a student. He had spent his vacation in mathematical pursuits, inspired by the community of Babbage, Bromhead, and the other members of the Analytical Society. And he had contributed original mathematical work to be presented to the leading scientific society in the kingdom. Now he could head back to Cambridge and prepare for the culminating Senate House exams and whatever lay beyond.

SCOTTISH CONNECTIONS

Something would be missing from Herschel's final year at Cambridge: one of his closest friends was not returning to the university. Though James Grahame was not part of the Analytical Society and was unique among Herschel's close friends in having no great interest in mathematics, his friendship was an important part of Herschel's early life. Whereas Herschel's friendships with Babbage and Bromhead were catalyzed by a shared interest in analysis, his close friendship with Grahame, which endured throughout both of their lives, was a matter of similar natures and shared experiences. And like his early mathematical training, this relationship came about by way of Scotland.

Politically unified with its southern neighbor only since 1701, Scotland had a tradition of independence in its educational and legal systems, and mathematical practices there were less strongly influenced by the legacy of Newton. The history of the Scottish royal succession meant French influence was stronger as well. Since James II of England (who was also James VII of Scotland) had gone into exile in 1688 as Britain's last Catholic king, his descendants in France had claimed a right to the British throne. Support in Scotland endured for this Stuart monarchy and went hand-in-hand with greater openness to French influences.

Edinburgh and Glasgow had by the early 1800s become important educational centers, providing training in mathematics as well as law and theology and remaining more connected than England to developments in France. The Scottish natural philosopher John Playfair, for instance, was an accomplished mathematician who reviewed Laplace's *Mécanique céleste* for the *Edinburgh Review* and, like Thomas Young, bemoaned the lack scholars in Britain who could understand it. It is no coincidence that Herschel's mathematical tutor was from Scotland and returned to Edinburgh once he had prepared Herschel for Cambridge. More than likely William, who corresponded extensively with Playfair, specifically sought out a Scottish tutor because it was only in Scotland that one could find the requisite mathematical expertise.

If Herschel's educational links with Scotland were already forged, his personal connections with the country were strengthened the summer after his first year at Cambridge. That summer, Herschel traveled with his parents as part of an extensive trip throughout northern England and Scotland, where William received honors and oversaw the erection of one of his telescopes. The Herschels had a family tradition of extensive travel throughout the kingdom. Besides the trip to Paris at the age of ten, there were journeys through the English countryside where John saw firsthand the birth pangs of the Industrial Revolution, a transition that would make Britain an imperial and commercial power during his lifetime. The summer before he left home for Cambridge, for instance, John and his family traveled to the manufacturing centers of Derby, Manchester, and Gretton as part of an extended trip to the Lake District. John's meticulous journal entries from this trip reveal an interest in manufacturing, including careful diagrams of the silk and cotton mills of Derby and detailed descriptions of the cloth factories of Leeds.[30]

FIGURE 1.2: James Grahame. From Grahame, *History of the United States.*

The summer after Herschel's first year in Cambridge, the family's travels took them to Birmingham, where the inventor James Watt (1736–1816) greeted his old friend William and gave them a tour of his foundry. The younger Herschel was immensely impressed with Watt's factory and felt it was "the source & fountain of every improvement in machinery which has displayed the power & ingenuity of man." Though he did not realize it at the time, the steam engines that were transforming the landscape were also creating both the fortunes that would allow a generation of amateurs to devote themselves to following William's lead in studying the universe and the tools and techniques to construct larger and larger telescopes to do so.[31]

In Birmingham, the Herschels' paths crossed for the first time with Grahame's family. Robert Grahame (1759–1851), a Glasgow lawyer and town magistrate, was also visiting Watt's foundry along with his wife and daughter. The Grahames and Herschels immediately connected, and William decided their trip would be extended to include a stay at Grahame's home of Whitehall, outside Glasgow. Here, Herschel recorded, they found themselves "instantly at home." The friendliness of the Grahames and the hospitality of Whitehall quickly put the normally reserved Herschel at ease. There was in their home, according to him, "none of the ceremonious politeness which torments the host" or "oppresses the guest" but rather only "an easy & most amiable hospitality."[32]

Part of the appeal of Whitehall was the fast friendship Herschel immediately formed with Robert's son James, a "dark-haired lantern-jawed slim young man" at this time two years' Herschel's senior. Within a week, Grahame was rousing Herschel out of bed by 7 a.m. so they could spend mornings clambering over the huge Forth and Clyde aqueduct outside Glasgow (and being berated as trespassers for their trouble), evenings dining with young ladies, and the days between in each other's company. Grahame was a welcome companion, as the Glasgow native helped Herschel avoid the more tedious aspects of the Herschels' visit. One evening Herschel accompanied his father and "80 or 90 Gentlemen" for a public dinner in William's honor. As dinner stretched on with interminable toasts and speeches, Grahame, sitting beside Herschel, proposed "leaving this Astronomical party & going in search of the stars themselves." This meant the ladies, who had dined at another location, and the two friends slipped away to join them.[33]

When it was finally time for the Herschels to leave Glasgow, Grahame accompanied them for a day on their return journey, traveling with them several miles along the Clyde. When Grahame finally departed, Herschel recorded in his journal that he was "a young man who (if I have anything of the prophet about me) will one day be mentioned as the ornament of his nation." Herschel may have overestimated the verdict of history, but Grahame would go on to practice law in Glasgow, write a book on the history of the United States, and compose tracts on the abolition of slavery. It was through Grahame that Herschel would eventually be induced to study law, a decision with important implications for his eventual views on science. Perhaps more importantly though, decades later it would be Grahame who introduced Herschel to his future wife, Margaret Brodie Stewart. The correspondence that Herschel and Grahame began that summer lasted for the rest of their lives.[34]

Grahame was at the time considering a career in law and told Herschel he hoped to eventually move to England. In the meantime, through their letters he asked Herschel for guidance on coming to Cambridge. Herschel no doubt regaled his friend with his growing passion for mathematics and the work of the Analytical Society. Though Grahame never had the desire or aptitude to pursue the subject, he bragged about attending the lectures of Playfair and assured his friend he was diligently studying algebra and geometry alongside natural and moral philosophy and music. The difficulty Grahame had locating up-to-date mathematical books illustrated that the challenge the Analytical Society faced securing textbooks extended to Scotland as well. In the fall of 1811, as Herschel began his third year as an undergraduate, Grahame enrolled as well, staying for the year.

The problems of evil and suffering continually arose in correspondence between the two young friends. Britain was going through a period of tumult, with economic disruptions caused by the war with France, the transformations of the industrial revolution, and a series of weak governments. It was a time of anxiety and uncertainty. The role of the laws of God and the laws of society may also have been on Herschel's mind as he considered whether to pursue a career as a clergyman. Grahame told Herschel of riding with the families of soldiers and officers traveling home by stagecoach from Cambridge to Scotland when they came across survivors from a stagecoach accident. "I trembled and felt sick at the sight and could not help asking myself with a feeling not very far from indignation, *What in God's name is the use of this misery?*" Herschel's friend would witness a great deal more suffering in his life, as he worked as a lawyer and fought (and sometimes failed) to keep his clients from the gallows. The role of law in preserving order, and what this meant for natural laws and the order of the universe, would soon come to dominate Herschel's thoughts—though he would ultimately apply himself not to the laws of society but the laws of nature.[35]

FINAL EXAMS

Herschel did not have much time to miss Grahame during his last year, as in a student's final year at Cambridge preparation for the Senate House exams became all-consuming. During a student's final term, no lectures were required, as time was theoretically devoted solely to cramming for the Tripos (though students were forbidden to work with a private tutor after their sixth term).[36] Herschel remained ambivalent toward the exam. On the one hand, he was confident it would cover little of interest or importance. On the other, success on the exams was necessary for social advancement. Students who scored top marks were more likely to obtain the highly competitive fellowships that provided semipermanent positions and continued lodging in the colleges. A fellowship meant a degree of financial independence and the opportunity to remain at Cambridge teaching and pursuing advanced studies. In addition, Herschel had a family legacy riding on his shoulders. Everyone no doubt assumed the son of William Herschel would perform at the top of his class.

Despite this pressure, Herschel was not satisfied spending his final term simply cramming. The Analytical Society had work to do. He and Babbage had decided sharing mathematical papers with other interested students at their society meetings was not enough to affect any real change at the university or beyond. An analytic revolution required those papers be turned into public offerings. In the midst of his preparations for the Senate House exams, Herschel and Babbage agreed they would build on the work of the Analytical Society the previous year by publishing a volume of memoirs. This way they could establish what they hoped would be the first of an annual series, a fledgling journal for their young society.

The *Memoirs of the Analytical Society* was the continuation of Babbage's original joke. Like those who argued that the truth of scripture was sufficient without accompanying explication or commentary, Herschel and Babbage decided the best way to promote the study of analysis against the darkness of British mathematical ignorance was likewise to exhibit it without commentary. The memoirs of their society would consist of papers showing forth the truth of analysis, giving the world the good news of deliverance from geometrical confusion. The papers were to be contributed by various members of the society. As it turned out, however, the rest of the society was not equal to this plan. When the first (and only) volume of the memoirs was published, it included just three long papers: two written by Herschel and one by Babbage.

Before this work became a reality, Herschel sat for the long-awaited exams. No one was surprised or disappointed in his performance. Herschel dutifully performed the mathematical exercises, which took place over four days, and finished in the top place as senior wrangler. Two weeks later, he competed for the Smith's Prize, getting top marks there as well. This performance meant he was virtually guaranteed a Cambridge fellowship. It also meant Herschel had been able to master the curriculum at Cambridge without breaking a mental sweat, taking top marks in the exams and at the same time pursuing the mathematical work that was his passion.

Herschel's friends were relieved he had not proven to be his own worst enemy on the exams, as Babbage proved to be. Despite his confidence in Herschel's performance, Grahame was worried from experience that Herschel's anxiety, which had at times disabled his friend, would be difficult to overcome. Herschel's former tutor Rogers offered his congratulations with a half-rebuke, writing his pupil that he should be proud of his success on the exams, "with which you seem so well disposed to quarrel." Those close to Herschel had been worried not that he would fail to gain academic success but that his frustrations with Cambridge would get in the way. Instead, Herschel found an outlet for his energies in the Analytical Society.[37]

By February of 1813, as Napoleon's power was beginning to collapse in Europe, Herschel was graduated and back home at Slough. The victories and frustrations of Cambridge were for the time behind him. Now he could focus on what he saw as the true fruit of his university experience. Far from a simple pastime among friends, the Analytical Society had become the catalyst for Herschel's vision of a British mathematical revolution. His father's telescopes waited, but Herschel had no time for astronomy. Herschel's analytical reform needed to be continued and extended far beyond Cambridge, and the first step was to promote the gospel of analysis through the publication of the society's *Memoirs*.

CHAPTER 2

A BRANCH OF ELEGANT LITERATURE

There is little or no taste for these things afloat—The maths are not here as on the continent considered as a branch of elegant literature.

—JOHN HERSCHEL TO EDWARD BROMHEAD, NOVEMBER 1813

London was in an uproar. Amid the continuing threat of Napoleonic France and the economic depression caused by Britain's Continental blockade, the United Kingdom had hardly any remaining allies left in Europe and faced growing discontent at home. Riots and uprisings swept the countryside, while authorities in the cities kept a close eye on seditious groups attempting to bring the liberating ideals of the French Revolution to British soil. Into this social powder keg the assassination of the British prime minister, Spencer Perceval (1762–1812), was tossed like a match.

In early evening of 11 May 1812, the prime minister of the United Kingdom was murdered in the Halls of Parliament in broad daylight. Perceval's assassin, a disgruntled Liverpool businessman, made no attempt to escape, confident his actions had been justified. As word of the deed spread, the public reaction was not one of dismay but largely—and horrifyingly for the members of Parliament barricaded behind the doors of Westminster—delight. Perceval's strict maritime embargos, including his support for abolition of the slave trade, were depriving merchants of profits and driving up prices for everyone. When a carriage arrived at Westminster to take Perceval's killer to prison, it was swarmed by a cheering mob trying to shake the assassin's hand. For a few breathless hours, it looked as though the situation would spiral out of control.

Guards were called out and patrolled the streets to keep peace. In at least one town, a band paraded in celebration of the prime minister's death.[1]

The tumult surrounding the prime minister's assassination was a dramatic episode in a long period of transition from which Great Britain would emerge as a global superpower and nascent empire. But that outcome was uncertain in 1812. At the time, a much larger concern was whether Britain would survive at all. For almost the entirety of John Herschel's life, the country had been at war with France, and recently that war had been going very badly. At home, the widely despised George IV ruled as regent in his father's stead. The mental illness of George III deprived the kingdom of a popular and respected ruler and Herschel's father of a patron who had supported his astronomical work and the construction of his huge instruments. Herschel graduated Cambridge during a period of significant social, political, and economic unrest.

At the center of the crisis of 1812 was the character of Perceval himself. He had been chosen prime minister in one of George III's final lucid acts and governed with a calm efficiency. Though many of his policies created powerful enemies, including the prince regent himself, Perceval relied on the legal training he had pursued, like many politicians before him, after graduating from Cambridge. For centuries, law had been one of the three traditional professions, along with medicine and theology. But the increase of industrialization across England in the late 1700s made legislation more essential than ever. Lawyers grew rapidly in numbers and wealth, and their profession began to shape society in new ways.

When Herschel left Cambridge to return to Slough, he found himself in Perceval's shadow. The prime minister had used his training in law in his attempt to reform society, but that attempt had foundered on the realities of politics and human nature, leaving behind a mourning wife and children. Against the backdrop of social uncertainty and the example of Spencer Perceval, Herschel tried to balance his own idealism with practicality. Law might be a means of reforming society, but what if the nature of society itself made this impossible? Mathematics seemed to point not only toward deeper laws than those governing society but, more importantly, offered a model of clear thought and certainty in a chaotic world.

"What should such a poor sniveling democratic dog do in this autocratic world?" Herschel asked upon his return to Slough. He was beginning a work he hoped would not only transform mathematical practice in Britain but establish a new "philosophy of invention" through clarity of symbolism and form. At the same time, he tried to navigate life beyond Cambridge, enveloped in a "cloud of engagements, dinners, parties, &c&c." Herschel's family wealth meant he had no immediate concern about employment, which was fortunate, as there were few if any opportunities to turn mathematics into a livelihood, yet he chafed at life in Slough, within sight of Windsor Castle and within reach of London society. "For my own part, were it not for family connexion," he complained to Cambridge friends, "I should never more approach the place." The only thing needed to mix in the superficial world of Regency society, he confided, was a "polished surface."[2]

A polished surface: Herschel may have been playing a word game here. His father was known for his polished surfaces, the mirrors created for his huge reflecting telescopes. Now that Herschel had returned from Cambridge, William wanted to teach his son to construct, maintain, and use these mirrors to take William's place at the telescope. William's great surveys of the sky were incomplete, and the aging astronomer hoped his son would continue his work. Herschel was frustrated with the society around him at Slough, but part of that included an awareness of the superficialities in his father's work as well.

William had a veneer of mathematical knowledge but was praised largely for the power of his instruments, not his theoretical insights. His son was uninterested in pursuing astronomical observations or constructing the tools and mirrors that made those observations possible, even less so if the results were primarily social: staying connected to the British royal family and selling telescopes to royalty throughout Europe. Observations were meaningless without tools to structure them into scientific knowledge, and those tools, Herschel was coming to believe, were mathematical analysis. In Herschel's mind, his task after Cambridge was clear: helping the Analytical Society transition from an informal group of students to a recognized philosophical force. Only when this failed would Herschel decide to follow the path of Spencer Perceval and take up the study of a different kind of law. And when that too fell apart, Herschel would be forced to reevaluate his entire approach to the laws of nature.

THE *MEMOIRS* OF HERSCHEL AND BABBAGE

As an undergraduate, Herschel had written several short papers presented at the monthly meetings of the Analytical Society. Now, with Herschel graduated and Babbage finishing his last year at Cambridge, Herschel began to expand two of these into articles to be published in the society's forthcoming *Memoirs*. If the society was to take its place among the learned societies of Britain, the group needed to make their work public. New societies were springing up all over Britain during this period, helping to move the production of knowledge from sources of credibility due to birth, wealth, or social standing to a new kind of rhetorical,

46

symbolic, and social authority—that of the scholarly community. The *Memoirs*, Herschel explained, would "serve to announce the existence of a society whose principal object is to lend all the assistance in their power to the improvement of the analytical methods in general" and especially in Britain, where such methods had for "too long been neglected."[3]

Herschel planned his first contribution to be a paper on trigonometric series, which he began writing only a month after his graduation. The paper expanded on the shorter pieces he had published in Nicholson's journal, developing methods for representing trigonometric functions as summations of algebraic series. The work was, he told Babbage, filled with "beautiful infinite products" and the sort of "strange oddities & inconceivable theorems" of which he was so fond. Formalism was primary, application secondary. Babbage, who was supervising the printing in Cambridge, had his doubts about this approach. After examining the printer's proofs, Babbage wondered if anyone other than he and Herschel would be able to understand them.[4]

As this first paper worked its way through the press, Herschel returned to Laplace's *Mécanique céleste*, the book he had quizzed his tutor on before arriving at Cambridge. Laplace's work represented the summit of analysis as applied to the physical universe, and the work inspired Herschel to write a second paper, this one on the topic of equations of differences. His work on this paper continued for the remainder of the summer. At the same time, much to the annoyance of the printer, he continued to add notes to his first.[5]

Herschel's first paper for the *Memoirs* was influenced by the work of the Cambridge mathematician Woodhouse, but his second paper showed him turning more toward the mathematicians of the Continent. Despite being in dialogue with Laplace's work though, there was no discussion of physical applications for the mathematical apparatus Herschel created, equations which at times stretched to such length on the page that the text had to be printed in landscape orientation to encompass them.

There was another reason to submit a second paper for the *Memoirs*: despite initial enthusiasm for the project, contributions from colleagues in Cambridge were not forthcoming. As far as publishable work, the Analytical Society was apparently a society of two. Luckily, there was a way to hide this. Early volumes of the Royal Society's *Transactions* carried no author names. Similarly, Herschel told Babbage, it would "carry a better appearance" if their volume was anonymous. Not crediting specific authors would make the *Memoirs* an offering of the entire society rather than a patchwork of individual contributions. It would also hide the fact that Herschel and Babbage were the sole contributing authors.[6]

In a sense, the *Memoirs* represented the capstone of Herschel's undergraduate experience, a sort of unofficial thesis exhibiting the extracurricular mathematical work he had pursued during his time at the university. But beneath the pages of equations, what, exactly, was the central message of the society? Babbage wanted to publish the papers without explanatory introduction. Herschel disagreed: "Place yourself in the situation of one of our readers (who possibly may be more numerous than you think). Would you not naturally say, Who are these people? Are we to have a series of volumes like the present? What are their resources, their views, their expectations? If we pass over all notice of this, consider how open we are to ridicule in the character of a few unknown Quixotic individuals who take upon themselves to enlighten the world."[7] Babbage consented, and he and Herschel authored a lengthy preface they presented to the society for approval.

In their preface, Herschel and Babbage illustrated the benefits of analysis with a long discourse on its nature and history. Analysis, they explained, was the process of pursuing long "trains of reasoning" that, due to their "length and intricacy," were impossible for human reason to follow without the structure of analytical expression itself. The strength of analysis was concise, clear, and symmetrical notation. Unlike a word or expression, a symbol in analysis "can neither convey, nor excite any idea foreign to its original definition." This made it possible for each step in an analytical process to be translated into common language, leaving the mathematician's memory unburdened of "all the load of the previous steps." Analytical expression could thus "condense pages into lines, and volumes into pages . . . shortening the road to discovery, and preserving the mind unfatigued."[8]

For Herschel and Babbage, analysis illuminated the process of thought itself. Understanding how the mind arrives at new insights through logical relationships could be harnessed for the process of invention. The logical steps of analysis forced the mathematician to observe the operations of the mind in the very process of mathematical discovery. It allowed an almost atomized view of this process, in which the mind retained connections with ideas derived "from reason rather than perception" while building toward new insights. Analysis was not simply a tool for mathematical progress; it was a window into the mental processes behind discovery.[9]

Clarity of notation was central to this power. For analysis to be effective, brevity and elegance of notation needed to be preserved and cultivated. In a passage excised (perhaps wisely) from their preface before printing, Herschel and Babbage castigated other British mathematicians—Edward Warring (1734–1798), Lucasian professor of mathematics at Cambridge, and William Emerson (1701–1782), author of a popular mathematical textbook using the fluxional calculus—for "exhausting whole alphabets in endless calculations, without the slightest regard to order, economy or symmetry." Only care and uniformity in notation opened the door to new mathematical insights. The mode of expressing a function by a single letter, for instance, made possible entire new fields of mathematics.[10]

This emphasis on notational clarity ran through both of Herschel's papers in the *Memoirs* and emphasized departure from the geometrical reasoning that had come before. In his second paper, for example, Herschel stressed the importance of his notation $\sin^{-1}$ rather than arc(sin) as "indispensably necessary" because it represented trigonometric functions as "mere numbers, functions of the number x." Putting restrictions on functions, such as forbidding imaginary or negative values because it was impossible to understand them physically or geometrically, was "worse than inappropriate." Mathematics could not be constrained to concepts that were graphically or geometrically clear. For centuries, spatial reasoning had provided the rational foundation for mathematics, but Herschel argued this was no longer necessary: it was enough that mathematics be logically or formally coherent and consistent. Though this conceptual transition had already played out on the Continent, in their *Memoirs* Herschel and Babbage were bringing it to British soil.[11]

Many conservatives feared importing French science would also import French revolutionary ideals, so the two Cambridge discontents had to tread carefully. In their preface, they portrayed their work not so much as *replacing* British mathematics as *reconnecting* it with Continental mathematics. British mathematics had fallen behind, certainly, but not because Newton's calculus was flawed. It simply had to be reimported "with nearly a century of foreign improvement." The Analytical Society would render Newton's calculus "once more indigenous" among British mathematicians. English pride was at stake: for too long, important advances in mathematics had come from abroad. The *Memoirs* would be an important first step in regaining mathematical preeminence.[12]

Bringing this analytical revolution to the physical page required a small-scale revolution in printing technology. The expressions Herschel used in his papers pushed the limits of typography and required careful oversight of the printing process, as any error would ruin the clarity that was the strength of the new method. Herschel had run into this issue with the publication of his first paper in the *Philosophical Transactions*, complaining to Babbage that it had been "translated into unintelligible nonsense by their confounded printers." To prevent this happening with the *Memoirs*, Herschel gave detailed directions for printing. The printer, for example, was instructed to always encroach on the margin when necessary so that equations would never be broken between lines, even if it meant pages had to be printed in landscape orientation. Clarity also extended to grouping equations. Even at "the hazard of leaving half a page blank," Herschel insisted, sets of connected equations must appear on the page together.[13]

Typographical innovations included more than just text placement. Typesetters had to replicate the subscripts and superscripts Herschel used for indices in his complex mathematical expressions, as well as parentheses, brackets, and radicals. Troublesome indexes had to be carefully positioned relative their variables: "not on top of the variable or on a line with it," Herschel spelled out carefully, "but off to the upper left." At one point printing was held up due to lack of numbers small enough for subscripts and superscripts. Herschel and Babbage were not simply creating a new form of mathematical expression in Britain; they were inventing the means of translating it to the printed page.[14]

Their diligence and careful instruction paid off. Whether anyone beside Babbage and perhaps their society colleague Bromhead appreciated the actual content of Herschel's papers, they at the very least were recognized as triumphs in mathematical typography, proving that the society's mathematical formalism could be successfully translated into print. After viewing the proofs, Herschel told Babbage he had never seen "anything so beautifully executed," and the other members agreed. The work that would become the first (and only) volume of the Analytical Society's *Memoirs* was printed with a run of 250 copies in late November 1813. The society itself ordered 100 copies, and Herschel and Babbage divided the rest between themselves. But even as the *Memoirs* were completed, the society was dissolving, with many of its members graduating and departing Cambridge. Herschel's hope was that as the group dispersed, the *Memoirs* would carry forward their mathematical revolution. He realized that to effect change in the British mathematical community, the society would need to find others who shared their passion beyond Cambridge, especially in London.[15]

Some of the members of the society, however, thought the work had gone too far. Herschel and Babbage had composed a collection of treatises as abstract as they were forebodingly complex. Even among mathematicians, few could follow or decipher the work or understand its implications. For casual readers, it was difficult if not impossible to see what use this new mathematics entailed. In essence, the volume was little more than a showcase for the analytical aptitude of the two young mathematicians. If the *Memoirs* were meant to light a fire for analysis throughout Britain, they failed.

Herschel quickly realized this. He wrote to Bromhead, now pursuing law in London, "the fire of enthusiasm spreads only where it meets with inflammable matter to receive & cherish it," but such a response was not to be found in Britain. Few were "disposed to enter heart & soul into a task of such gigantic labour, and such diminutive reward." Mathematics in Britain led to "no public distinctions" and afforded "no prospect of pecuniary reward." As far as publications, if a book went "one step beyond the comprehension of *Elementary readers*," Herschel complained, it was "a dead weight & a loss to its author." It was perhaps unsurprising the public took no notice of the *Memoirs*, but its authors were especially frustrated that neither did the scientific or literary community. Their work was not reviewed by a single British periodical.[16]

Things did not improve when Herschel returned to Cambridge the following fall. He had been granted a fellowship in the spring after his graduation, which provided lodgings at St John's and required residency during the terms, when he was expected to tutor students. But his new role made involvement in a society of radical mathematicians problematic. There was no talk of reestablishing the society's monthly meetings, and despite earlier arrangements for payment of the *Memoirs*, there was the awkward matter of collecting funds. Herschel did not want to pressure a group of undergraduates to pay for the publication of what was basically his own work, believing he and Babbage should bear the remaining costs themselves. "I do not I confess at all repent the publication," Herschel confided to Babbage, "but I believe every one else must look upon it as a very mad scheme."[17]

It was in this atmosphere of self-doubt and disagreement that Babbage chose to defend his radical proposition for his act of disputation and was quickly failed by the outraged examiner. Any hope that Herschel's mathematical partner in the *Memoirs* would distinguish himself on the Senate House exams evaporated. Babbage's behavior was embarrassing for Herschel, for their mathematical project, and for the reputation of the Analytical Society. When Herschel confronted him, they argued, and Babbage promptly left Cambridge for the Christmas break without saying goodbye, leaving Herschel so upset that he made up his mind not to write. Their partnership seemed to have met an abrupt and ignominious end.[18]

Herschel learned from his mistakes. If the Bible Society that triggered Babbage's original notion had been correct about providing the gospel without commentary, Herschel had provided the wrong gospel. Examples of analysis had not converted the mathematically unenlightened. His work could not function as the pure, undiluted truth of the new mathematics. Instead, the society needed to provide the gospel of analysis in a way that all could understand. Herschel's old tutor Rogers pointed this out when he received his copy of the *Memoirs*, emphasizing the importance of providing an introduction and overview of this new form of mathematics, rather than examples of advanced analysis in action. The society's next reforming step would be providing this necessary background by translating the French textbook that had converted its members to analysis to begin with.[19]

Herschel's second insight from the failure of the *Memoirs* was that a society to promote analysis was still a worthwhile endeavor—if it could be correctly executed. Outside Cambridge, the place to establish such a society was London. Ultimately, the successor to the Analytical Society would not emerge among the gentlemen enthusiasts and natural philosophers of the Royal Society but instead among bankers, insurance brokers, and amateur astronomers of London's rising middle class. In the Astronomical Society of London, Herschel would eventually find the ideal balance between analytical techniques and their application. The Astronomical Society, besides threatening the hegemony of the Royal Society, would eventually carry out Herschel's mathematical reform by transforming the practice not of pure mathematics but of astronomy itself.

THE DECISION FOR LAW

As Herschel completed his papers for the *Memoirs*, the formation of the Astronomical Society was still years away, and the question of a potential vocation loomed on his horizon. The most viable option was a career as a clergyman in the Church of England, and his family and friends assumed he would follow this path. Herschel's close friend and fellow Cambridge graduate Richard Jones (1790–1855), for instance, followed the trajectory of many graduates when he took a position at a parish and pursued his economic studies while living the quiet life of a country parson. Even Herschel's colleagues who remained at the university as fellows fulfilled ecclesiastical duties (which often amounted to no more than an occasional sermon) in surrounding churches or chapels.

Herschel was distinctly resistant to this assumed path. It did not matter, as his father would soon argue, that in practice the role of a minister in the Church of England allowed for exceedingly vague theological convictions. At this point in his life, Herschel felt that organized religion, of either the high or low church variety, was itself the problem. Evangelical enthusiasm, which had intensified in England since the late 1700s, came in for his particular scorn. "Let them . . . pour forth the bitter vials of religious controversy," he wrote as he was considering his plans after graduation, "so long as we are not obliged to take a part in the game." At its best Herschel felt a religious vocation would make him complicit in a legally sanctioned system of spiritual governance. He could not "turn Methodist or high churchman, or even religionist in the received sense of the word. Heaven forbid!" Despite his protestations though, as long as he was attending Cambridge the church remained an assumed vocational goal.[20]

Herschel began seriously considering a new possibility regarding a professional vocation when his family received Robert Grahame, William's longtime Glasgow friend and father of Herschel's close friend James, as a visitor to Slough during the months Herschel was working on the *Memoirs*. The elder Grahame had a significant influence on the young mathematician, who recounted that simply being around Grahame made one a better person. Grahame no doubt updated Herschel on his son's career as a lawyer in Scotland, and their conversations during the several days of Grahame's visit gave Herschel a new vision for a path to apply his passion for reason and clarity of expression to society. By the time Grahame departed for Scotland, Herschel had decided to pursue a career in law.

Besides the church and medicine, law was the third of the ancient professions for which a Cambridge education provided the foundation. Two of Herschel's closest friends, Bromhead and James Grahame, were already studying or practicing it. And though there was no immediate financial need for Herschel to earn an independent income, his conversations with Robert Grahame no doubt recalled the long discussions Herschel had had with James about reforming society's ills when they were students together. Herschel became convinced that, just as James was doing his part through law in Scotland, he could do the same in England—and more effectively than he could through a career in the church.

Being a lawyer could also be quite lucrative, as the careers of Grahame and Bromhead ultimately proved. Already James Grahame was making a significant living, telling Herschel he had earned "a monstrous sum for a lawyer so young in his profession." Lawyers prospered in the young British empire as their fortunes grew alongside the growth of industry. Between 1800 and 1850, the expansion of the canal system and railway network brought a vigorous land market and corresponding litigations. The number of lawyers significantly outpaced the general increase in population, from about 250 in 1780, to more than 800 practicing lawyers in 1810, to more than 3,000 by the middle of the century, the vast majority of whom were based in London. Lawyers led the push for professionalization, and there was a growing sense within the field of service to society, an ideal James Grahame embodied in his own work. It

FIGURE 2.1: British society and law established on natural laws. Title page, *The Real or Constitutional House That Jack Built* (London: J. Asperne, 1819).

was a booming market for students of law, and with his freshly minted Cambridge degree, his attention to detail, and his logical eye, Herschel was well suited to be part of this growing profession.[21]

Despite these enticements, Herschel's father opposed his son's decision. William found John's own objections to the church unconvincing. The two had a long discussion on the topic during a rainy carriage ride in early November 1813. William could not understand his son's logic. As a parson, William argued, John would have more than enough time to pursue mathematics and would contribute to the moral health of his parish even if he did not agree with every particular of church doctrine. If this seemed hypocritical to John (and it did), William pointed out that as a lawyer he would be in an even worse position. In court, John would be a *professional* hypocrite, required to argue for his clients even if he knew they were in the wrong. The discussion turned into an argument, and their trip ended without agreement. A few days later a conciliatory letter from his mother reached Herschel in London, along with an enclosed note from William, but Herschel remained resolute.[22]

James Grahame, on the other hand, was thrilled to hear of his friend's decision. "I have never known a man better qualified to excel as a Barrister than you," he told Herschel. "If you will devote yourself to your profession you will . . . attain an honorable independence and distinguished rank and above all the power of being eminently useful." Law required tenacity, capacity for recall, and diligent attention to detail, logic, and rigor—all things Grahame knew Herschel had in abundance.

Herschel himself eventually compared the work of the lawyer in the courtroom to that of the natural philosopher investigating nature. Experiments were ways to "cross-examine" the witness of nature and untangle the simple principles behind apparently complex behavior.[23] Similarly, a relatively simple law or legal code could become quite complex in application. As Grahame explained, success in legal studies did not mean possessing a huge body of facts. Rather, "the learning of an enlightened lawyer" came from understanding the legal structures upon which cases were decided. A lawyer's work consisted "in a great measure in drawing distinctions and discovering analogies." Grahame's advice would indeed become central to Herschel's practice—but not, ultimately, in the courtroom.[24]

In mathematical analysis, Herschel had found the power of clarity, order, and conciseness as tools of thought and methods of discovery. Law was an opportunity to apply similar tools to human conduct and society. The political and legal order of Britain were seen as flowing from the natural law-like order of the universe itself. (In France, disregard of this order had resulted in revolution and anarchy.) Herschel's desire to establish himself in a profession that would let him usefully serve and beneficially reform society fit well with the law-like order he had found in analysis. In addition, his decision came at a time when his mathematical hopes were at a low ebb with the failure of the *Memoirs* and his falling out with Babbage.

It was time, Herschel felt, to head to London.

LONDON

By 1813 London was well on its way to becoming the largest metropolis in the world. It had overtaken Amsterdam as the world's leading financial center late in the previous century. By the turn of the century, the population of the city had reached more than a million, a number that would nearly double in the next four decades, driven by rising life expectancy, increased crop yields, and industrialization. London was the warehouse of the world, with millions of pounds of freight passing through each year. Five years after the assassination of Perceval, it was estimated that at any one time 3,000 barges, 3,000 passenger vessels,

40,000 laborers and 12,000 revenue officers were employed handling the trade passing through the 4 miles of dockyards along the Thames. London was a rising commercial powerhouse and at the forefront of the world's first wave of major urbanization.[25] After the 1813 Christmas holiday, instead of returning to Cambridge, Herschel took up lodgings in the city and enrolled at Lincoln's Inn, one of London's four ancient law schools.

Once settled, Herschel began his transition to a professional career, dutifully beginning work in the office of a Mr. Saunders. He was, he told his father, "sinking from the theory to the practice—from the spirit to the forms of the law." Legal cases came to Saunders's office for decision and afterward were passed to a group of students who would read them and discuss. By February, Herschel was spending five or six hours a day in the office. "It is not wonderfully amusing, but still not nearly so dry as I had had it represented." When offered another tutoring position at St John's, he declined, saying that his studies would not allow him time to teach as well.[26]

Grahame warned Herschel that to succeed he needed to devote himself entirely to legal studies: "Consider the law a rich and disquieted mistress that will not be content with divided homage." It soon became clear, however, that divided homage was all Herschel would be able to offer. His journals continued to fill with mathematical notes, despite his hours at Saunders's law office, and his rooms in London put him distractingly close to the centers of British scientific society. He soon began attending weekly meetings of the Royal Society, held a short walk from Lincoln's Inn in the imposing Somerset House overlooking the Thames.[27]

Herschel had been elected a fellow of the Royal Society the previous May. For membership, one needed the recommendation of a current fellow, which Herschel's father was eager to provide for his mathematical son. Elected at the age of twenty-one, Herschel was among the youngest fellows yet admitted to the venerable society. But by the time of his arrival in London, the Royal Society had become less an institution for scientific research than a status symbol, a way for cultured gentlemen to dabble in natural philosophy and mingle with others of their class. Reforming the Royal Society would eventually become a long, painful, and—at least initially—unsuccessful battle for Herschel and his peers.

If Herschel found the Royal Society moribund, the Royal Institution, located on Albemarle Street a half hour's walk from Somerset House, was by contrast where spectacular and obvious progress in science was being made nearly daily. The Royal Institution had been founded in 1799 to support public engagement in science—a somewhat more egalitarian answer to the Royal Society's elitism—and had quickly become the center of a growing interest in chemistry, providing regular lectures open to the public. Herschel soon began attending institute lectures on chemistry and mineralogy.

Chemistry was on its way to becoming the most important and successful science since Newton's gravitational physics. Following the work of Joseph Priestly and Antoine Lavoisier, many in the first two decades of the 1800s looked to chemistry for the promise of significant scientific and industrial progress. At the Royal Institution, advanced equipment like the giant voltaic pile of Humphry Davy (1778–1829) functioned as the inverse of William Herschel's telescopes, allowing a view into the nature of matter as it broke substances into their constituent elements and brought with it a cascade of new discoveries. Chemistry, with its impressive equipment and explosive demonstrations, captured the public imagination as a source of both power and insight into the secrets of nature.[28]

The growth of chemistry went hand-in-hand with the growth of mining and the analysis of minerals. Travel abroad and excavations at home brought strange rocks and crystals into the laboratory, and whereas anatomical or botanical specimens might be dissected for study, chemistry provided the tools necessary to separate, investigate, and classify the components of minerals. Herschel was first exposed to mineralogy at Cambridge through the lectures of Edward Daniel Clarke (1796–1822), a clergyman and naturalist who traveled extensively throughout Scandinavia, Russia, and the Middle East in the late 1700s and was made professor of mineralogy at Cambridge during Herschel's student years. Clarke had built up a huge mineral collection (he was said to have brought home eight hundred samples from Siberia alone) and was a popular lecturer with infectious enthusiasm. Herschel began helping Clarke analyze specimens, work he continued in London.[29]

Chemical and mineralogical pursuits, Herschel admitted, did "not look very lawyerlike." At Royal Society meetings he heard of the latest scientific discoveries firsthand, while at the Royal Institution he witnessed the chemical demonstrations and apparatus that made the discoveries possible. At one society meeting Davy read a paper on his discovery of iodine, and the very next week William Hyde Wollaston (1766–1828), a chemist who had become wealthy through his method of processing platinum ore, brought along a sample of the new element in a glass tube. After his student years pursuing abstract mathematical analysis, Herschel now found himself in the company of Britain's leading natural philosophers investigating the building blocks of nature in ways that could be seen and touched. Cambridge had seemed a mathematical backwater; in London groundbreaking physical discoveries were being made all around him. Within weeks, Herschel had drawn up a list of chemical equipment to purchase and began keeping an experimental notebook recording not legal precedents but nature's laws. By March 1814 his notebook included results of almost daily experiments. As far as law, he assured his father, "I still believe even *that* profession is not all-engrossing."[30]

Herschel was not alone in this pursuit. Babbage, now graduated from Cambridge and independently wealthy from his father's income, came to London as well, where the two friends reconciled and began conducting experiments together. For the two mathematicians, there were significant parallels between chemistry and mathematics. Chemical analysis broke minerals into their constituent parts, using solvents to dissolve samples ("wet analysis") or blowpipes to heat them ("dry analysis")—but when Babbage asked Herschel whether he was "analyzing in the moist or the dry way," he meant whether he was working on chemical or mathematical analysis.

For chemical analysis to be successful, its procedure needed to be as systematic, detailed, and rigorous as mathematical analysis. Each step was recorded in detail so other investigators could follow or re-create the process. Just as mathematical analysis depended on clear notation, chemical analysis depended on clear nomenclature. An accurate and logical system of chemical names was a concern from the very start of Herschel's chemical work. On the other hand, mathematical analysis never resulted in the violent explosions that Herschel's chemical experiments sometimes did. And unlike mathematics, the results of chemical analysis, when not lost to explosion or fire, could be carefully weighed, described, and even tasted. A particular result from an experiment in the spring of 1814, for instance, was found to be "excessively *acrid* taking away sensation *& colour* from tongue and roof of mouth."[31]

In the midst of this chemical experimentation and legal studies, Herschel had still not given up on his mathematical reform, despite the failure of the *Memoirs*. In early 1814 he completed a second mathematical paper for the *Philosophical Transactions*, another attempt to bring the work of the Analytical Society to a broader audience. The *Memoirs* had failed at least in part because of their inaccessibility, so in this paper Herschel took pains to explain his formalism and even provided problems and examples, including applications to geometry. Despite his turn to chemical analysis (and ostensibly law), he retained his drive to reform British mathematics from the inside out.[32]

Yet after a season in London and in spite of his close proximity to the Royal Society and the Royal Institution, by mid-1814 Herschel was ready for another change. Another nineteenth-century chemist had claimed

London was "worth one's while to see once; but the most disagreeable place on earth for one of a *contemplative* turn to reside in constantly." Herschel's experience seemed to bear this out; in the years to come he would be at his most productive and happiest away from the metropolis. Now, after half a year shuttling between London and Slough, with his notebooks filling with almost daily chemical experiments, Herschel wanted to return to the slower and more contemplative atmosphere of Cambridge. As a fellow of St John's, he still had rooms available at the college, and so as fall approached, he made plans to return.[33]

RETURNING TO CAMBRIDGE

In September 1814 Herschel fulfilled a prophecy he had made earlier to Babbage that they would turn their colleges into laboratories by setting up his chemical apparatus in his rooms at St John's. His supplies included forty pounds of mercury, new scales and weights, and a detonating tube. As soon as he arrived, he set to work re-creating the experiments of John Davy (1790–1868), Humphry's brother and assistant at the Royal Institution, and performing mineralogical analyses of local soils and waters. His experiments continued throughout the fall term, ceasing only for the Christmas holidays. He resumed his work with Clarke as well, attending his lectures and analyzing more samples from his extensive collection. Herschel had now, he informed Babbage, become "half a mineralogist."

All this no doubt felt a world away from the courts and law offices of London. Yet Herschel assured Babbage he had returned to Cambridge "solely for the purpose of reading law, and am really studying it with avidity." Grahame, ever watchful from Scotland, was not impressed: "You seem to be always chasing something and always despising it, praising something else and neglecting it," he chided his friend. According to Grahame, Herschel needed to get out of the laboratory and into the courts. Instead, Herschel applied for the position of professor of chemistry at Cambridge.[34]

Herschel's chance for a professional vocation in science came with the unexpected death of Smithson Tennant (1761–1815), who was killed in a traveling accident after Herschel's first term back at Cambridge. A founding member of the Geological Society, Tennant had been awarded the Royal Society's prestigious Copley Medal in 1804 for his chemical work and had been made professor of chemistry at Cambridge in 1813. Herschel had attended Tennant's lectures along with Clarke's as a student and spent time working to re-create some of his demonstrations. A professorship was one of the few positions in the entire kingdom that provided the freedom to devote oneself entirely to natural philosophy, so Tennant's death, tragic as it was, offered Herschel an opportunity for a professional livelihood. Cambridge had only a handful of such positions, which were not awarded by any particular college but rather by the university itself. Each was established by an endowment and carried the (often neglected) responsibility of lecturing. Herschel had not been able to find a way to make a living as a mathematician; perhaps he would have better luck as a chemist.[35]

Herschel decided to put his name forward to fill Tennant's chair. The move was an implicit decision to give up law: if he was awarded the chair, he could not continue his legal studies. As it turned out, the position had already been spoken for. It was awarded the chemist James Cumming (1777–1861), who held it until just before his death in 1860. Cumming, to Herschel's frustration, was relatively unknown as a natural philosopher and not yet even a member of the Royal Society. His selection over Herschel meant that despite Herschel's newfound passion for chemistry, he was still seen primarily as a mathematician. "If Tennant had lived a twelvemonth or 2 years longer," he told Babbage, "I should in all probability have been his successor."[36]

Whatever Herschel still claimed about his law career at this point, it was becoming clear that his overriding passion was analysis, both wet and dry. Yet as his experiments began to turn toward optics and the nature of light, the balance he was attempting to maintain at Cambridge started falling apart. As a tutor, he was required to assist with examining students, a process he found an onerous drain on his time. "At present I am plunged in all the horrors of Examination," he told Babbage. He was also responsible for writing problems for the exams. "I put it off till the last moment," he confessed, "and then did them all in one morning, so that one half of them are *cram* & the other, wrong." His law studies and experimental pursuits, he felt, were becoming too much to balance alongside his college responsibilities.[37]

Herschel was twenty-three years old and felt pulled in a dozen directions. He had poured himself into the Analytical Society's *Memoirs*, only to see it fail to make any impact on the mathematical community. He had fought his parents over his desire to pursue law, only to leave London for Cambridge. He had begun extensive experimentation in chemistry but had failed to secure a professorship. In March of 1815, Herschel abruptly broke off experimentation in the midst of an analysis of feldspar, left his laboratory, and returned to the city. But the change of scenery was not enough to revive his flagging legal studies. Distance from Cambridge simply shifted his analysis from wet back to dry. His mathematical work resumed, but he was reaching a point of crisis. Throughout the summer, Herschel found himself spiraling toward nervous exhaustion.

The antidote was enforced vacation. During the early 1800s and into the Victorian era, the mind was thought to manage the body's stresses much as a governor on a steam engine controlled pressure inside a boiler. Too much stress on the mind endangered the health of the entire body. Herschel's doctor ordered him to the seaside resort of Brighton in the south of England, a few hours' journey from London. Herschel obeyed, but instead of relaxing and enjoying the sites or the ocean airs, he complained about not having his chemical equipment. "If you have got any numbers you want calculated to a hundred decimals," Herschel wrote Babbage in preparation for his trip, "give them me." In Brighton, Herschel roamed the seaside, feeling he was "the idlest mortal that ever idled away an idle life" and pestering Babbage about borrowed chemical equipment he wanted returned when he was back in Cambridge. Instead of prescribed idleness, his work in Brighton became a third paper on mathematical analysis for the *Philosophical Transactions*.

In early drafts of this paper, Herschel compared analysis to a garden that needed "perpetual pruning and culture" or quickly became "disfigured with shaggy formulae," perhaps echoing his own mental state as he tried to balance his mathematical pursuits with law and chemistry. In another draft, analysis became a game of chess, a means of simplifying thought that always "at one glance" presented the state of the game. The mathematician, like a chess player, was "left at liberty to direct the whole vigour of his intellect *forwards*, and to concentrate his efforts upon the real difficulties he has to overcome." For Herschel, analysis remained a means of transforming thought—and a pathway for clear and lucid cognition.[38]

Despite his restlessness, Herschel's Brighton stay was a turning point. The stress that had precipitated it forced Herschel to acknowledge the truth. "Brighton has done me good," he admitted to Babbage, "but I shall not be able for a long time, if ever, to resume my professional studies in London." Instead, he planned to return to Cambridge once again, where he would support himself by tutoring, continue his chemical experiments, and drop the pretense of law. He was concerned about his health, planning to take on only a handful of students and split his time between Slough and Cambridge until, "by dint of regularity & care," he acquired "a sturdiness of constitution [which] will enable me to be of some use in the world." It was possible he would return to law one day, he thought, but for now it was a "lost game."[39]

For Babbage, Herschel's second return to Cambridge meant another attempt at their efforts of mathematical reform. As a tutor, Herschel was again positioned to influence mathematical pedagogy. "What a glorious opportunity you have of spreading the true faith," Babbage crowed, dubbing his friend "the Apostle of Analysis" who would "purge away the diagrams which like cobwebs have obstructed . . . progress in the paths of truth." Yet by the end of another term, Herschel was once more fed up with the process of tutoring and exams. Not only had the process taken time away from his other pursuits, he felt by the end of the term that he had failed to make even one of his students understand his analytical approach.[40]

By the year's end, despite his mathematical publications Herschel had yet to find a means of moving the reforming program of the Analytical Society forward. Now it seemed he was not cut out for tutoring in analysis either. With these failed approaches in mind, Herschel returned to the Analytical Society's original inspiration. He and his colleagues had proclaimed the gospel of analysis through their own work, and it had failed. It was time to go back to the original source of their inspiration and bring *that* to the mathematically unenlightened of Britain.

TRANSLATING LACROIX

In their enthusiasm for importing analysis to British soil, the Analytical Society had put the metaphorical cart before the horse. The reason the mathematical tools developed on the Continent and perfected in France had made so little impression in Britain, despite the work of Herschel and his friends, was the lack of any comprehensive English textbook to explain them. What was needed, as Herschel's former tutor had urged, was a text that could provide an introduction and overview of the new mathematics. Without this, Herschel realized, their own work in the field remained largely incomprehensible to their countrymen.

Herschel and Babbage had been aware of this need since the publication of their *Memoirs*. The other members of the society agreed. Bromhead, for instance, admitted that the *Memoirs* had been "too profound to do . . . any good" because "not one mathematician in 10^∞ can understand them," maintaining that "the true faith will never flourish till a book has been published in *English* . . . in a *compact* & *tangible* shape." Scottish mathematicians agreed. One complained that there was "not a book in the English language from which anything like a tolerable knowledge of the fluxional or differential calculus, in its present improved state, can be obtained." Playfair in particular felt that Britain's low mathematical competence was not due to the difficulty of the works of Continental mathematicians but rather "from want of knowing the principles and the methods which they take for granted as known to every mathematical reader."[41]

There was no introductory textbook in English that would provide the needed background, but there was one in French: the *Traité élémentaire du calcul différentiel et du calcul intégral* by Lacroix. This was the book Babbage had purchased for such an exorbitant cost during his student days at Cambridge and which had provided the kernel for his original Bible Society spoof. Babbage had even started a translation of the book during his student days but had never completed it. Lacroix's text was the true gospel of the Analytical Society. With their decision at the start of 1816 to translate his textbook, the Analytical Society came full circle. To translate it now, Babbage and Herschel would be assisted by George Peacock, one of the original members of the society. Babbage would translate the first third of Lacroix's text, Peacock the second, and Herschel the third.[42]

Besides Herschel, Babbage, and Peacock, there were other colleagues at Cambridge still interested in the cause of analysis. Gwatkin, another member of the society, was "function mad," and their friend William Whewell was beginning to read "foreign mathematics" and *"useful things."* The three translators hoped this was only the tip of the iceberg of interest their work would generate at the university. As Peacock looked for a printer, Herschel, mindful of the complications that had arisen in funding the *Memoirs*, insisted on working out all the financial details beforehand. They settled on a printer who would produce one thousand copies, with each book priced at fifteen shillings, making it easily affordable for most students and an order of magnitude less expensive than the original French version Babbage had purchased. The book would turn no profit, but an inexpensive and comprehensive analytical textbook would be available in Britain for the first time.[43]

Immediately prior to beginning work on the translation, the reforming momentum of the Analytical Society resumed somewhat as the paper that Herschel had begun working on in Brighton was published in the *Philosophical Transactions*, alongside two others by Bromhead and Babbage. Three papers by members of the Analytical Society appearing together in the Royal Society journal was a step in the right direction, reminding the wider community that work on the topic was still progressing. But a deeper reform was at play in the Lacroix translation itself as Herschel and his colleagues introduced a change beyond simply language. In his original text, Lacroix used the theory of limits as the basis of his formulation of calculus. In this approach, the derivative of a function is expressed in terms of the values of a function as it approaches a specific point. However, Herschel and the Analytical Society preferred the older and less geometrical method of Lagrange, which established calculus on the basis of the Taylor series, an infinite sum that expresses the value of a function's derivative at any point. Lagrange's method, they wrote in the translation's preface, was "more correct and natural." This change placed the formal foundations of analysis in Britain on a different footing than contemporary developments in France. In Lagrange's version of analysis, mathematical functions do not themselves stand for or measure rates of change, limits, or any other physical conceptual basis. They are "of themselves, as meaningless as the rules of grammar, or those of a game." By revising Lacroix's text so that it was based on Lagrange's formalism, Herschel and his colleagues were stripping mathematics down to something as clear and precise as the rules of chess—with no necessary basis or connection to the physical world at all.[44]

In addition to these foundational revisions, Herschel and the other translators also significantly extended the original book's scope. If the Analytical Society was presenting the gospel of analysis through the scripture of Lacroix, it was still proving to be their own version of that gospel, with their own commentary. Lacroix's original work included an appendix of nearly one hundred pages, and Herschel replaced this with an even longer one of his own. Herschel, Babbage, and Peacock also added a series of notes extending over another 130 pages to provide additional proofs or explain passages that might be unclear to English readers.

Herschel wanted the translation to be useful for teaching, replacing the outdated Cambridge mathematical texts that were still being used to prepare for the Senate House exams. But it needed to be more than this. Herschel hoped that readers of the translation, even if they were motivated by cramming, "shall not be able to look into it without picking up something better, & getting a tinge of the true faith." The textbook would communicate as many Continental innovations as possible, "secrets which have hitherto been contraband." Finally, the translation continued Herschel's crusade for mathematical clarity when it came to symbolism itself. The method of separating symbols of operation from those of quantity, for instance, would itself "sell the book."[45]

Their translation was finished by the end of 1816, and Herschel, Babbage, and Peacock each began planning their own companion volumes to provide even more examples and further extend the text. Unlike the *Memoirs*, which had garnered the Analytical Society neither sales nor readers, the Lacroix translation was a measured success. It answered the need for a comprehensive textbook on the new mathematics, and the first copies sold quickly. Moreover, it provided a foundation for curricular reform at Cambridge. With a textbook in hand, members of the Analytical Society could begin shaping questions posed on the Senate House exams. Peacock, who was to be an exam moderator that year, pledged to use the new symbols for questions he posed. "The hole [*sic*] of St John's," he informed Herschel, "resounds with daily altercations of the d'eists & the dottites." This was progress. Five years earlier no one outside the Analytical Society had cared.[46]

On the other hand, the furious reactions that Peacock's exam questions elicited showed the resistance Herschel and his friends still faced. The other exam moderators, Peacock told Herschel, "were very angry & threatened to protest against analytic & French mathematics." Only the success of his students, Peacock believed, saved him from actual disciplinary action. Analysis was problematic, its opponents maintained, not only because of its French ties but because it subverted the order of mathematics and thus the relationship of the mind with truth. Traditional mathematics was founded on its connection to the physical world. Analysis, especially the Lagrangian version that Herschel and the others had used in their translation, broke this connection. What the Analytical Society saw as the strength of the new mathematics, conservatives saw as its danger—political as well as moral.[47]

Even Herschel's friend Whewell, who was originally sympathetic to the society's aims, thought it would have been better if Peacock had posed problems so difficult that the new methods of analysis provided the only way to solve them. This would have shown the utility of analysis, Whewell explained to Herschel, and made the examinees "thank you for your way of doing them." Until they saw its use, students and fellows of Cambridge would not believe its value, even in translation. Peacock had succeeded only in causing an uproar, Whewell said; he had "stripped his analysis of its applications & turned it naked among them."[48]

THE RELUCTANT ASTRONOMER

As Herschel's program of analytical reform continued in Cambridge, his father's astronomical career was drawing to a close. By now William's days as an active observer were behind him, but there were plenty of remaining projects he hoped to see completed. The gigantic forty-foot telescope, still the largest in the world, sat unused and in need of service. So did the more functional twenty-foot, which required new mirrors. The catalogs resulting from William's sweeps of the night sky contained thousands of double stars and nebulae no one else had ever seen. Those sweeps needed to be repeated. Reobserving double stars to see how their components changed would provide confirmation that Newton's law of gravity extended beyond the solar system; reobserving nebulae could verify his theories on stellar evolution. William's health would not allow him to complete this work himself, and Caroline, who had performed the calculations that transformed his observations into data, had long since left England for her childhood home of Hannover.[49]

In the fall of 1816, as the Lacroix translation neared completion,

William traveled with John to the coastal town of Dawlish, in Devon, on the south coast of England. Dawlish was near the village of Torquay, where Babbage was staying with his own ill father, and Herschel looked forward to roaming the countryside with his friend looking for mineralogical specimens, likely interested in the area's formations of slate, limestone, and sandstone that had drawn the attention of other geologists. For William, though, the trip had an ulterior motive. His mathematical hopes for his son had been realized and exceeded. John had the mathematical tools that could enable him to use William's observations and provide a mathematical theory of the sidereal universe just as Laplace had provided for the solar system. And John would have the most powerful instruments in the world at his disposal, as well as training in how to use them—if only he could be convinced to accept this inheritance.[50]

The old astronomer knew his time was limited. So far, the younger Herschel had shown little interest in taking up his father's work. Though he had peered through his father's instruments and watched at his aunt's elbow as she reduced observational data, John considered himself a mathematician and chemist, not an astronomer. Yet if he did not take up William's work, there was no one else. No one beside William knew the methods for working the huge telescopes. And despite the extent of his work, astronomy still remained focused on things within the solar system. William needed his son to continue the sidereal revolution he and Caroline had begun. If he did not, observations of nebulae and double stars would be lost for years, perhaps decades, until another observer with comparable instruments returned to them. Without William's systematic approach, though, any future observer would be starting from scratch.

In Dawlish father and son stayed with the son of William Watson, the natural philosopher who had first connected the elder Herschel with the Royal Society when he was still a musician in Bath with an interest in building telescopes. In discussions on the long ride to Devonshire or walking along Dawlish Water where it tumbled over artificial waterfalls to the sea, William, now approaching his seventy-eighth birthday, made his case. He emphasized that a career devoted to astronomy would provide John freedom to pursue his other interests and that John's financial inheritance made working at a profession unnecessary.

Herschel talked things over with Babbage. Perhaps anxious to escape his father's pressure campaign, Herschel hastened from Dawlish to Torquay. As he and Babbage rambled the district, "climbing the rocks . . . and talking over analytics," their discussions reawakened all the enthusiasm he had felt when they had first formed the Analytical Society.

No doubt they also discussed their fathers' wishes for the both of them and their plans and hopes for the future. Babbage commiserated with Herschel's dilemma—to remain at Cambridge or return to Slough and take up his father's work. Herschel was also, like Babbage, concerned for his father's health. The ostensible reason for the retreat to Dawlish was a long illness William had recently recovered from, and Herschel noted a marked decrease in his father's strength. He had to face the realization that there was a limited window in which William would be able to pass on his knowledge. By the end of their time in Devonshire, Herschel had agreed to become his father's apprentice.[51]

Back at Slough, William immediately created a list of projects his son needed to begin, which included learning how to polish the telescope mirrors, rebuilding the framework of the twenty-foot telescope, and resuming sweeps. William's telescopes were unique in that instead of using lenses to gather light—as did refracting telescopes like the one originally used by Galileo and those that still formed the primary tools of observational astronomy—William cast huge mirrors for his telescopes. William's reflecting telescopes allowed him greater light-gathering power and higher magnifications than any other telescopes of the time. Because the entire manufacturing process had been created and refined by William and his brother Alexander, Herschel had to learn not only his father's observational method but also the craft knowledge to create, shape, and polish the mirrors. Herschel's chemical notebooks soon carried details on experiments with speculum, the metal used for his father's mirrors. It quickly became clear it would be necessary to cast entirely new mirrors for the twenty-foot before sweeps started up again. In the meantime, Herschel became familiar with significant objects and the basics of observing with a smaller seven-foot reflecting telescope. On 12 October 1816, Herschel began a new notebook dedicated to astronomical observations. His career as an astronomer had begun.[52]

Herschel still had misgivings. To observe for long nights at the telescope meant giving up his rooms at Cambridge and leaving the university, likely for good this time. He was nonetheless allowed to keep his college fellowship despite no longer being in residence, which speaks to the influence of the Herschel name and the perceived importance of the work he would be doing. Still, Herschel was heartbroken to leave. "I always used to abuse Cambridge," Herschel admitted to Babbage, "but, upon my soul, now that I am about to leave it, my heart dies within me." The departure affected him so much he felt too depressed to do any mathematics. On top of that, he had found a huge mistake in one of his *Memoirs* papers and wondered if he should buy up all the remaining copies to remove them from circulation.[53]

Yet even as Herschel took his place at the telescope, circumstances brought about the chance for a new mathematical work. In Herschel's third *Philosophical Transactions* paper, published in 1814, he had referenced in passing the work of William Spence (1777–1815), a mathematician working in relative obscurity in the mercantile town of Greenock who had published a single short mathematical treatise in 1809 that few outside the Analytical Society could have understood or appreciated. At Spence's death, a friend who had read Herschel's paper realized Herschel was likely the only person in England who could edit and publish Spence's unfinished work. Accordingly, in late 1816 Herschel was surprised by the unexpected delivery to Slough of a trunk of Spence's manuscripts.

Though Herschel had never met Spence and knew little of his life, here was the chance to champion the work of another British analyst. Over the next several months, Herschel diligently prepared and arranged a volume of Spence's essays, which were ultimately published in 1819. Unfortunately, besides copies Herschel sent to friends, the rest of the books disappeared into the stock room of a Scottish publisher and were not seen again for twenty years. Another avenue of Herschel's analytical reform had again foundered on the rocks of general indifference and incomprehension.[54]

Yet Spence's essays encouraged Herschel to keep the idea of an analytical reform alive. In the midst of editing Spence's work, Herschel wrote to Babbage about reviving the Analytical Society, this time in London, with a broader base of mathematicians. For this attempt, Herschel felt the key to success would be to do things in "a more moderate way." Instead of "the nonsense of having weekly or monthly meetings, an annual or quarterly dinner in Town will do just as well." Herschel was learning to adapt his reform to the London professional class. This new instantiation of the society never materialized, but Herschel's purposed arrangement would eventually become the model for another reforming society: the Astronomical Society of London.[55]

Herschel was also at work following up on the Lacroix translation with his own original textbook on algebra. The failure of the *Memoirs* and the comparative success of the Lacroix text had driven home the point that reform in mathematics would depend on textbooks. This would go hand-in-hand with the kinds of new problems that could be posed on the Cambridge exams. Peacock, for instance, wanted an algebra text that promoted their new mathematical methods without appearing forbiddingly difficult. Herschel's name alone, his friends hoped, could overcome any opposition to it being adopted at the university.

Herschel worked on his textbook, which he said would have a plan different from any other work on algebra, throughout the summer while he also worked to get mirrors fit for the twenty-foot telescope. William was anxious for new ones, and in May John traveled to London to have two cast at his father's direction. Casting mirrors was one thing; polishing them was another. Herschel trained for hours on his father's polishing apparatus. It was tricky business: at times a mirror would be made fit by polishing for high magnification, but then, in an attempt to finish it, the polish was ruined so badly Herschel could barely see his face in the reflection. Actual observing went on slowly.[56]

Unlike his first faltering steps in astronomy, Herschel's *Algebra* never bore fruit. All that remains are a few manuscript pages and a portion of his drafted table of contents, which show that in the book, as he had with his other mathematical works, he planned to emphasize the role of analysis in thought. The first article of the first chapter, for instance, was outlined as "saving of attention & memory, [and] the effect this has on judgement." Herschel emphasized algebra's power to reason about relationship between unknown quantities. From the portions remaining, it is clear Herschel planned to set out the fundamentals in keeping with his analytical program but written as accessibly as possible, beginning with simple steps of solving algebraic equations. With its emphasis on the abstract role of relations in algebraic symbolism, Herschel's unfinished text was a forerunner of the more radically formalistic algebra George Peacock would publish over a decade later.

Meanwhile, despite its setbacks the previous year, by spring of 1818 the analytical reform at Cambridge seemed to have gained some ground. Members of the original society were now tutoring students, and Peacock could report that "the spirit of analysis is spreading very rapidly amongst the young men of the college." At the same time, Herschel was asked to contribute articles on mathematics to a new encyclopedia under preparation by the Scottish natural philosopher David Brewster. These articles would allow Herschel to outline his view of mathematics to an even wider audience. In the midst of this, Herschel's *Algebra* lost its urgency and was soon abandoned.[57]

TOWARD APPLICATION

From his Cambridge graduation, Herschel's work had not strayed from promoting a reforming view of mathematics centered on clear nomenclature and abstract reasoning. Through his foray into law, his back and forth between Cambridge and London, his chemical experiments, and his ultimate decision to take on his father's work, mathematical analysis remained a central pursuit. He published the *Memoirs*, worked with his colleagues on translating Lacroix, rescued Spence's work from obscurity, and tried his hand at his own textbook. Herschel wanted to use a reformed mathematics to reform *thought*, not to connect those mathematics to practical applications in the laboratory or at the telescope. If Herschel was anything during this period, it was a mathematician—and one who gloried in the abstract nature of mathematics.

Herschel had told Babbage not long after their student days that he "existed more in theory than in practise" and "made the beautiful & the abstract his cynosure in mockery of base utility." But by 1818 this idealism of abstraction was beginning to change. In his article on mathematics for Brewster's encyclopedia, perhaps considering the years he had devoted to its study, Herschel insisted that all the work put into analysis, "all the lives drawn out in abstract speculation," could not be considered wasted, because "the very next physical problem which presents itself for examination" could depend on the tools of analysis for its explanation. The justification of his mathematical labors might be in its connection to the physical world after all. Herschel's views were changing as he began a new line of experiments, and this transformation took on a new urgency when he and Babbage decided to travel to the Continent.[58]

Herschel had worked to bring French mathematics to Britain; now it was time to travel to France and meet those mathematicians in person. In Paris, Herschel came in contact with those whose work he had been promoting for so long and witnessed how natural philosophy and mathematics could work together within a strong institutional structure and with generous governmental support. Herschel would soon position himself at the center of London's scientific community, and the first step was to reestablish those international relationships strained by the long wars with France.

CHAPTER 3

REVOLT OF THE ASTRONOMERS

An abstract truth, however, is of no country.

—JOHN HERSCHEL, "MATHEMATICS,"
EDINBURGH ENCYCLOPEDIA, 1818

Shortly after his return from Egypt with Napoleon's army, the French chemist Claude Louis Berthollet purchased a country home at Arcueil, a small village outside Paris. The fine château was situated amid gardens and vineyards and provided a retreat from the bustle and responsibilities of the capital that William Herschel had enjoyed on his visit to Paris in 1802. Soon after, Berthollet's friend and colleague Laplace purchased a house on adjoining grounds. Together, Laplace and Berthollet assembled around them an informal society of French savants whose careers and research flourished under Napoleon's generous patronage. Arcueil, with its long crushed-gravel walks, rippling brook, and picturesque Roman ruins, became the fashionable center of French scientific society.[1]

For much of John Herschel's life, this scientific idyll was out of reach. For as long as Herschel could remember, apart from the brief pause of the Peace of Amiens, his country and France had been at war. It was a conflict with significant personal costs for the Herschels. Though England was spared invasion, the German kingdom of Hannover, home to much of Herschel's extended family, bore the brunt of invasion. Throughout Herschel's childhood, his father received despairing letters from his younger brother Dietrich as their home suffered under the occupation of first French and then Prussian forces.[2]

Yet in the years since Herschel left Cambridge, the political landscape of Europe had been transformed. With the final defeat of Napoleon at Waterloo in 1815, commerce and travel between Britain and the Continent resumed. Under the restored Bourbon monarchy, France entered a period of peace, economic stability, and industrialization. In January 1819, Herschel and Babbage set off for what would be Herschel's first trip outside Britain since his childhood visit, intent on meeting the French mathematical and scientific figures whose work they had been promulgating in England. In Paris they would witness firsthand the significant institutional support for science that continued to flourish even after Napoleon's fall from power.

To the young mathematicians, the French scientific enterprise contrasted sharply with that of England. What Herschel saw in Paris underscored his belief in the need for scientific reform in his own country and led to a significant shift in his own scientific practice. The French Revolution had given natural philosophy a place of privilege as a tool for constructing an enlightened, secular state. In contrast, science in England was seen as a way to stabilize society, preserve the balance of power, and cultivate virtue—if it could be rid of corruption and its institutions reformed. With Napoleon's ascendency, institutions like the Institut de France (eventually to become again the Académie des Sciences) and the Bureau des Longitudes that provided salaried positions for men of science were strengthened and brought into service of the war effort. Natural philosophers like Laplace and Berthollet increased in political power and used their influence to promote their extensive program of mathematical physics. By 1819, despite the political turmoil of Napoleon's fall, the physical sciences in France were supported and professionalized to an extent far beyond in Britain.

Herschel's time in Paris and his conversations at Arcueil were a turning point for his early career. Besides meeting the savants whose papers he had been reading and whose work he had been following from afar, Herschel was exposed to a model of support of the sciences, particularly astronomy, that highlighted the morbidity of the Royal Society in London and influenced the shape his plans for reform would take in England. The results of this trip were immediate, with the establishment in London of a band of professionals who would take up the reforming task of the Analytical Society and directly challenge the authority and prestige of the Royal Society.

The conversations, demonstrations, and experiments Herschel participated in while in Paris also motivated a new focus in his own research: the application of mathematical analysis to the physical world. After Paris Herschel would no longer be content to pursue pure mathematics. He returned from France with a renewed determination toward a scientific life and a plan for applying mathematics to physical experimentation. The key to unlocking the mysteries of nature, he realized, was not mathematics alone but rather analysis applied to the frontiers of physics.

RETURNING TO PARIS

Herschel and Babbage had discussed plans to travel to the Continent for some time. Babbage proposed a trip as early as 1816 to buy books, though the two finally agreed they would leave after Christmas 1818 and accordingly departed in early January. They set off from London, spending an evening at the home of Joseph Banks, the Royal Society president, who likely shared with them accounts of his own attempts to support French science and maintain international scientific connections during the darkest days of the wars. A generation earlier, in 1793, Banks had been astonished that a certain French philosopher would not visit him to discuss natural philosophy. "I cannot conceive," Banks had thundered, "that any one would consider as a political necessity to debar me from the acquaintance of a Learned man because he is of a nation with which we are at war." Banks supported their trip now that Europe was at peace again, not suspecting that what Herschel and Babbage saw in Paris would influence them to undermine his authority over London's community of natural philosophers.[3]

Before departing London, Herschel sent off two papers to David Brewster in Scotland for inclusion in Brewster's new *Edinburgh Philosophical Review*. The papers provide a view of the state of Herschel's research on the eve of his departure. The first was a long article summarizing his chemical work on "hyposulphurous acid," a sulfur compound Herschel had isolated and for which he carefully recorded properties, characteristics, and assumed composition. Ultimately, this paper would be seminal for the invention of photography, as it was here that Herschel noted the ability of this acid to dissolve silver halide, the material eventually used to expose images on photographic plates. Herschel's acid became the photographer's hypo used to "fix" photographic plates, the process that for over a century was the basis for developing photographs.[4]

The second paper Herschel sent from London could be considered his final paper of pure analysis, emphasizing the utility of analysis's "simplicity of expression, and utility of transformation." Summarizing his work in the field so far, he referenced his Lacroix translation, which showed "the elementary simplicity and luminous elegance" of applying analysis to the method of differences, and his 1816 paper in the *Philosophical Transactions*, which he said extended his mathematical method and made it the basis of "a new mode of symbolic representation." In this summative paper for Brewster's journal, Herschel applied his program of mathematical analysis, still devoid of physical application, to summing specific types of algebraic series. As these papers made clear, on the eve of his first Paris trip, Herschel was still pursuing the dual tracks of mathematical and chemical analysis.[5]

From London, Herschel and Babbage took a coach to Dover and then sailed to Calais with an easy two-hour passage of the Channel. From Calais, Paris was a two-day ride by *diligence*, an enclosed coach shared with five other passengers. On their first night on the road, the two travelers arrived late to a poor inn where they found no dinner and had to make do with "a bottle of sour Bordeaux and some capital Brandy, which mixed together served us for tea"—at 1 a.m. But they were bound for the center of the scientific world and much more comfortable lodgings. They arrived on 16 January in Paris, where they found rooms "furnished with *blue velvet* chairs and a polished oaken floor" at the Hotel de Bruxelles on the Rue de Mail.[6]

Even more than London was to England, Paris was to France the center of the nation's social life and governance. Napoleon's centralizing programs had drawn the nation's administration and educational institutions to the capital, which in turn had gathered scientific practitioners from all over the country. Whereas in England, outside of London one might find isolated natural philosophers working in Cambridge, Edinburgh, or Manchester (or even, like Spence, doing mathematics in the relative obscurity of Greenock), in France the centripetal force of the Institut de France and the École Polytechnique had brought practically all French practitioners of the physical sciences to the capital. The scientific monopoly of Paris worked in the travelers' favor: all the important French natural philosophers and mathematicians could be easily found in one place.

If Paris was the social and political center of France, Laplace's scientific research program remained the thematic center of French science. The mathematical work that Babbage and Herschel had so closely followed and emulated from across the Channel had, in France, developed alongside a distinct approach to Newtonian physics shaped and directed by Laplace. The goal of the Laplacian program was to explain all physical phenomena in terms of attractive and repulsive forces modeled using the mathematical analysis so prized by the Analytical Society. Laplace's great synthesis, the *Mécanique céleste*, illustrated the power of this approach for explaining the structure of the solar system. Parisian natural philosophers pursued Laplace's program applied to heat, electromagnetics, sound, and light. Of these areas of research, Herschel became most interested in developments regarding light.

Herschel was especially eager to meet Jean-Baptiste Biot (1774–1862), a disciple of Laplace and professor of astronomy at the University of Paris who by this time was giving courses on light, sound, and magnetism based on his own research. Herschel had been closely following Biot's work from England, as his own experiments had recently transitioned from chemistry and mineralogy to optics—like chemistry, a vibrant subject in the midst of radical transformation and development. Though Biot and most of his French colleagues worked in terms of Newton's theory that light existed as tiny particles emitted from luminous objects, in 1801 the British experimenter and mathematician Thomas Young had reopened debate on the nature of light by demonstrating that it could interfere with itself like a wave. Things became more complex when the French engineer and mathematician Étienne-Louis Malus (1775–1812) discovered in 1809 an additional property of light, eventually known as polarization. Light could be polarized by its interaction with matter, opening a new avenue to explore the nature of light and potentially combine optics with mineralogy and chemistry. Biot's research prior to Herschel's arrival involved this critical intersection: the interaction of polarized light with matter. In a series of papers in 1812, Biot had applied Laplace's approach to light to successfully explain newly discovered optical phenomena, rejoicing that he had incorporated light into Laplace's "dream of a single broad explanation for physical phenomena."[7]

Herschel's first optical experiments involved measuring the refractive power of different fluids. In the laboratory in his rooms at Cambridge, Herschel made tables of the properties of different chemical solutions, listing their specific gravities, capillary action, viscosity, freezing points, boiling points, specific heats, expansibility, and compositions. From such careful tabulations, Herschel hoped to arrive at general principles connecting the physical properties of a substance to how light interacted with that substance. By 1818 his early experiments had evolved to include re-creating Brewster's investigations of the polarization of light through its refraction and reflection by certain materials. The summer before his Paris trip, Herschel's close friend William Whewell could write to him asking about his work and saying he had heard Herschel was "untwisting light like whipcord" and "cross examining every ray that passes within half a mile."[8]

When Herschel met Biot, the nature of light was still very much an open question. Newton's corpuscular theory could explain most optical phenomena, but the application of this theory remained problematic for certain situations. Polarization in particular posed a challenge. Before Malus's discovery, it was known that certain substances, such as the mineral called Iceland spar, created two images of any object viewed through it. Somehow, materials like Iceland spar broke incoming light into two rays, a phenomenon known as double refraction. When the light passed through the material, instead of simply being refracted in a single direction as in normal transparent matter, light was refracted in two directions, creating what was known as the ordinary ray and the extraordinary ray. Moreover, rotating a crystal of Iceland spar in certain directions made the second image disappear when passed through a second crystal. Double-refracting materials not only split light: they somehow gave light a directional property—the property eventually labeled polarization.

Malus's discovery that even normal objects such as glass could polarize light by reflection made its study even easier. Polarization seemed key to understanding both light and the nature of matter, providing a link between Herschel's previous chemical and mineralogical experiments and new optical research. Perhaps more importantly, investigation of polarization was especially suited for the mathematical tools Herschel had been cultivating since his student days. Biot's experiments attempted to reduce the relationship between polarization and double refraction to a single mathematical equation. His approach showed that optical research, and in particular polarization, was a chance to apply mathematical analysis to the nature of light itself.[9]

On their second day in Paris, as soon as they had recovered from their overland journey, Biot called on Herschel and Babbage at their hotel and escorted them to a meeting of the Académie des Sciences (which the first class or scientific portion of the Institut was renamed in 1816). The contrast between France's learned scientific society with London's Royal Society immediately struck the two travelers. Whereas the Royal Society was open to nearly anyone with the right connections and social status and had a membership that had now grown to nearly six hundred, the membership of the Académie was carefully managed. Limited to seventy-five, positions in the Académie were divided among scientific subjects with no more than six members assigned to each area of research. Rather than a society of gentlemen amateurs, the Académie remained a branch of the French government, with salaried positions and expectations for serious research. Though the Royal Society might be consulted by the government on occasion, in contrast to the Académie it had to find its own premises and support itself with subscriptions from its members. Other aspects, however, may have been more familiar to the travelers. As at the Royal Society, papers were read aloud at Académie meetings. One contemporary visitor commented that this made the meetings "as tedious as those of the Royal Society of London."[10]

Biot introduced Herschel and Babbage to all the leading figures of

French sciences and mathematics, many of whom had met Herschel's father when he had visited in 1802. At that time, William Herschel had been paid the significant honor of election to one of the eight *associés étrangers* in the first class of the Institut. Such positions were available to foreign natural philosophers only when one came open by death of a previous holder. Laplace, now seventy, remained the reigning patriarch of the French scientific community, his political savvy allowing him to retain his wealth and influence even after Napoleon's fall. Laplace hosted Babbage and Herschel at Arcueil, where they discussed the work of Herschel's father on nebulae. The French mathematician felt William's work on the development of stars from nebulous material supported his own dynamic view of the formation of the solar system and planets. Herschel had his doubts on this account but kept them to himself.[11]

Herschel and Babbage also met Siméon Denis Poisson (1781–1840), another disciple of Laplace, who applied mathematical analysis to the description of electromagnetism. There was also François Arago (1786–1853), another astronomer of the Paris Observatory who had traveled extensively with Biot and was applying mathematical analysis to problems in positional astronomy like the refraction of light due to Earth's atmosphere. There were mathematicians such as Adrien-Marie Legendre (1752–1833) and Joseph Fourier (1768–1830), who would apply mathematical analysis with great success to the conduction of heat through materials. And, of course, the English mathematicians were especially pleased to meet Lacroix himself, whose textbook had inspired their Analytical Society and whose work Herschel and Babbage had now translated and to which they were soon to contribute even more examples.

The first meeting of the Académie Herschel and Babbage attended was chaired by Georges Cuvier, the famous zoologist, naturalist, and paleontologist. The Swedish chemist Jacob Berzelius (1779–1848) and French chemist Louis Jacques Thénard (1777–1857) were in the audience, along with Joseph Louis Gay-Lussac (1778–1850), who had ascended with Biot four thousand meters in a hot air balloon to determine whether the strength of Earth's magnetic field varied with height. (They determined that it did not appreciably.) Each of these natural philosophers and mathematicians had achieved significant success and recognition in their respective fields. For Herschel, it was a heady experience to find himself in the company of so many whose work he had studied and admired from afar. They in turn knew him as the scion of William Herschel but also as an original contributor to their own mathematical program.[12]

Beside the chance to meet all the leading members of the French scientific community, Herschel was especially struck by the professionalism and the coordination of research of the Académie. Here was a structured hive of productivity, in which its members were financially supported and expected to do original research, with the understanding that such research would ultimately benefit the state. Biot, for instance, had conducted investigations on the origin of meteorites, the refractive index of gases at varying densities, global magnetism, the velocity of sound—using newly laid water mains in Paris for his tests—and the conductivity of metals in addition to the optical work Herschel was so eager to discuss. The French astronomer did not need to worry about another profession, such as the church or law, to support his scientific interests. In France natural philosophers were supported and celebrated, and Herschel's first Académie meeting felt a far cry indeed from the less productive and more haphazard pursuits of the Royal Society.

For the next three weeks, Herschel busied himself in a round of dinner invitations, experiments, and meetings. He attended more gatherings at the Académie and dined again at Arcueil. In the relaxed and informal surroundings of Laplace's villa, Herschel and Babbage compared notes with their French colleagues on approaches to mathematics, and Herschel began to feel he had neglected a major aspect of the utility of analysis in the pure mathematics he had pursued since Cambridge. This suspicion grew as he conducted experiments with Biot on polarization and discussed at length possible physical mechanisms and models to explain it.

While in Paris, Herschel also called on Arago at the Paris Observatory, where he learned more about the functioning of the Bureau des Longitudes. The bureau was the French answer to the British Board of Longitude, which had been established in 1714 to support British attempts at determining longitude at sea. The bureau, itself established in 1795, had been modeled on the Board of Longitude, but as the Académie outstripped the Royal Society, so the bureau was more professionalized and supported than the board. It consisted of two mathematicians and four astronomers, each of whom were paid a significant salary and viewed their research as a full-time occupation, not a sinecure or consulting appointment as it was for members of the Board of Longitude—something Herschel would soon learn to his frustration when he became involved with the board upon his return to England. The bureau met once a week, and its members, which included Laplace, "considered it their duty to devote their complete working day to the advancement of astronomical theory and the calculation of accurate astronomical tables."[13]

In short, Paris exposed Herschel to a new way of doing natural philosophy in terms of social structure, financial and governmental support, and experimental approach. The Herschel who left England at the beginning of 1819, pursuing wet analysis on one hand and dry on the other, was not the same Herschel who returned a month later. He had glimpsed the integration of mathematical and experimental work in the field of optics, and he had seen the functioning of a well-supported scientific society. Herschel came back to London with a transformed view of science and a new vision for its reform.

REFORMING ANALYSIS

Though Herschel published more articles on hydrosulphurous acid after his return from France, his experimental work now turned almost exclusively to optics. Immediately upon his return to Slough, Herschel picked up where his work in Paris with Biot had left off. He gathered specimens of double-refracting minerals to test Biot's optical theories and constructed an apparatus of two thin plates of material that he could rotate to create and adjust polarized light. He began closely examining substances, trying to understand the optical phenomena he observed in terms of the structure of the materials light passed through. This was more than simply a new avenue of experiment; his mathematical method was changing too.

Before his trip to Paris, Herschel's analysis had remained largely divorced from application. In the eyes of some of his colleagues, this pure abstraction contributed to the failure of his initial reforming attempts at Cambridge. Herschel had been content with showing the power of analysis to contribute new *mathematical* results. But in France he had seen firsthand analytical methods informing experimental method and the interpretation of results. When Herschel returned, his analysis became grounded in physical application. As Herschel would soon realize, polarization provided a way of mapping mathematical equations directly onto the physical world.[14]

Herschel's new attitude was immediately obvious to his closest mathematical collaborator. The abstract mathematical exercises that had previously filled his correspondence with Babbage were no longer enough for Herschel. Babbage complained that his friend had been converted to a focus on utility and told Herschel that he himself refused to worship that "golden calf." Babbage disagreed with Herschel's new approach, maintaining that each analyst should "be allowed to select those dishes most agreeable to his palate." In other words, Babbage would not let his own analytical pursuits be dictated by physical application, and he resented this new constraint in Herschel's work.[15]

In 1818, as he began his optical experiments, Herschel told Whewell, "It will be long before I can generalize sufficiently to *discover*, but if I only open my eyes I may *observe*." Though a classical inductive approach may

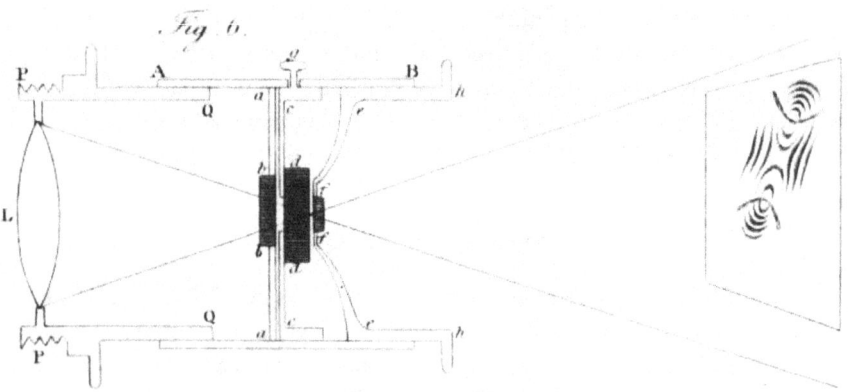

FIGURE 3.1: Diagram of Herschel's apparatus for projecting colored rings of polarized light. Herschel, "On the Action of Crystallized Bodies," plate V.

have been sufficient for his early chemical experiments, after Paris Herschel began to use mathematics to *shape* his observations. He had begun his optical work without any "definite or accurate ideas" on polarization. After working with Biot, though, he had a specific research direction: explaining the deviations between Biot's theory of polarization and what was actually observed. Passing polarized light through double-refracting materials resulted in rings of colors, beautiful and striking optical displays that provided a means of measuring optical quantities. Herschel felt he could arrive at the "general laws" that "regulate the action of crystalline bodies on light" and attempted to synthesize his experimental results into a broader theoretical framework that could explain double refraction and polarization. He was, he claimed, now working toward "a *general analytical theory* . . . for any law of double refraction whatever." For Herschel, mathematical analysis would no longer bear fruit in clarity of thought alone but instead take a central role in the creation of theory.[16]

The key to Herschel's approach was precise measurement of the colored rings created by passing polarized light through double-refracting materials. Biot had provided a quantified relationship between these rings and their widths for different angles of incidence of light and thicknesses of crystal. Herschel wanted to take this further and find a mathematical relationship that completely explained and described the rings. Measuring the rings provided an indirect means for determining the extent of double refraction in the crystal, which was especially useful when the angle between the ordinary ray caused by normal refraction and the extraordinary ray formed by double refraction within the crystal was too small to be measured directly. Herschel's method would allow optical properties to be determined for crystals too small or thin to otherwise measure.

His earlier work in mineralogy had already primed Herschel to consider directly measuring mathematical relationships in nature. In crystallography, the branch of mineralogy related to crystals, measuring the angles between faces of crystals as accurately as possible was essential to generalizing about their interior structure. In 1809 Herschel's friend William Wollaston had invented a reflective goniometer, a device that made these precise measurements possible. This invention, Herschel would later recall, "changed the face of minerology" and gave it "all the characters of one of the exact sciences." Biot's work helped do the same for polarization. In Paris Biot showed Herschel tables of data with expected results based on derived equations compared to measured experimental results. Like Wollaston's goniometer, analysis made mathematics itself a tool for probing the interior structure of matter.[17]

Herschel took this analytical turn even further. His indirect method of measuring double refraction via polarization allowed him not only to compare his measurements to the predicted mathematical results but, he felt, to directly *perceive* the mathematical relationships at play. His method enabled the creation of formulas based "not on a laborious, and sometimes deceptive discussion of tabulated measures, but on the actual form of the curves themselves." In Herschel's mind, this moved mathematical analysis from the creation of equations describing the behavior of light to experiments in which light directly *displayed* those equations, letting him copy the mathematical "outline fresh from nature." When Herschel passed light rays through his polarizing apparatus in his darkened laboratory, he was not simply building tables of measurements. He was seeing mathematics itself etched in patterns of light.[18]

Herschel presented this new work in a paper read to the Royal Society in December 1819, nearly a year after his return from France. In the paper, Herschel arrived at a mathematical expression that, he said, proved the axes of double refraction in the same crystal varied for different colors of light. For Herschel, this result was independent of whether light was assumed to be a wave or a particle. It was also independent of any physical assumptions regarding forces in the crystal's molecules acting on light rays. Though Biot and his colleagues still tended to interpret their results with the Laplacian paradigm of forces between particles, Herschel remained agnostic.[19]

The 1819 paper represents an important shift for Herschel. Before, Herschel had pursued mathematics as a pure mathematician, celebrating and promoting the formalism and clarity that analysis provided. It was a tool for thought, needing "no assistance from inductive observation, and very little from the evidence of our senses." Even when applied to the physical world, the mathematician reasoned on the products of analysis "as creatures of his own imagination." But in France, Herschel was exposed to mathematics as a tool for understanding the interaction of light with matter. Apart from the supplement of examples based on his Lacroix translation, Herschel's work now shifted to the application of analysis to physics. For Herschel, pure mathematics would never again hold the appeal it once had, naked and without grounding in physical application. In a very real sense, his work with the Analytical Society had come to an end. Soon, Herschel had the opportunity to apply analysis to another field, with a new group of colleagues. The analytical reform transitioned into an astronomical revolt.[20]

CALCULATIONAL ASTRONOMY

As Herschel was reevaluating his approach to mathematics, a London stockbroker named Francis Baily (1774–1844) was discovering a passion for mathematics applied to astronomy. Initially, calculating star positions as a pastime was the furthest thing from Baily's mind. Apprenticed to a mercantile firm at fourteen, by the age of twenty-one Baily had decided on the life of an explorer, sailing to the United States and passing two years in the wilds of North America. He overwintered on a boat trapped in ice on the Ohio River, met the American frontiersman Daniel Boone (1734–1820), and passed through Washington, DC, while the city was still being cleared from the surrounding forest. Baily returned to England in 1798 and, when he failed to raise funds for travels in Africa, instead turned his attention and energy to stockbroking and risk analysis, something with which he had gained personal experience on his travels. He quickly established a reputation for the accuracy of his actuarial computations, with publications that helped modernize the field of actuarial science. Yet when he presented a paper on the topic to the Royal Society, it was rejected.[21]

The Royal Society remained firmly under the autocratic rule of Joseph Banks, its president for more than forty years. Banks, who had traveled with Captain James Cook (1728–1779) to Brazil, Tahiti, and Australia and brought back the tens of thousands of botanical samples that helped make the Royal Botanic Gardens the greatest in the world, saw himself as the gatekeeper and guardian of gentlemanly scientific practice in Georgian London. In contrast to what Herschel had seen in France, the Royal Society remained a club dominated by aristocratic members who could participate in the trappings of natural philosophy without requirement to make scientific contributions. Natural history and botany were prioritized, often to the neglect of mathematics. Banks had already survived a mutiny of sorts over precisely this issue several years before, with mathematicians resigning in protest of what they saw as the exclusion of colleagues who lacked adequate social standing. This early reforming movement failed, and in the eyes of many the society had become increasingly moribund in the years since.[22]

Baily's interests turned toward applying mathematical theory to astronomy, and with the Royal Society's rejection of his paper he now had a chip on his shoulder against London's scientific elite. His passion for calculation and history led him to a project determining the precise date of an ancient solar eclipse, which in turn initiated a career in calculational astronomy. This was not the astronomy of telescopes and observing; instead, it was the meticulous compilation and calculation of star positions. Such "ledgers full of stars" served for navigation and timekeeping, the tools that let British ships more safely navigate the globe and decreased the investment risks that brokers like Baily spent so much time considering. Baily went on to publish several catalogs of star positions. Along the way, his spat with the Royal Society turned into a full-blown feud with London's scientific establishment.

Positional or calculational astronomy of the type Baily pursued had a long history in Britain. Though latitude on Earth's surface could be determined directly by measuring the altitude of stars, determining longitude was more complicated. For a seagoing mercantile society, it was nonetheless essential to have a reliable and accurate means of doing so. Since 1714 the Board of Longitude offered prizes to anyone able to solve the problem of determining longitude at sea and worked with the astronomer royal, the royal appointee who directed the Royal Observatory at Greenwich, to develop and test potential methods. Knowing the exact position of stars was a central component of this endeavor, and the astronomer royal was charged with creating the requisite star catalogs. Since 1767 part of this work had included publication of a *Nautical Almanac* to help sailors calculate longitude based on the distance of certain stars from the moon.

Measuring the positions of stars to create accurate catalogs was anything but a straightforward process. Observing was only the first step. Afterward came a rigorous process of reducing those observations: accounting for atmospheric refraction, the aberration of starlight, and subtle changes in Earth's orientation in space caused by the oscillation of its axis and precession. It was with these complex and time-consuming calculations, so removed from the observational astronomy William Herschel had pursued with his large telescopes, that Baily thrived. Though he conducted no observations himself, Baily's calculations made observations dating back decades and even centuries useful to contemporary astronomers. These catalogs were the backbone of positional astronomy. For people like Baily, tables of life annuities and catalogs of star positions were two sides of the same calculational coin: both documents transformed raw data into useful (and profitable) knowledge.

Baily soon came to realize this type of work was neither recognized nor appreciated at the Royal Society. Calculational astronomy was tedious labor, carrying with it no glamour of discovery. It was true that Banks and the Royal Society had embraced the work of William Herschel and had published John Herschel's papers on mathematical analysis, but the elder Herschel's work, despite its uniqueness, still fit into the mold of natural history: gathering observations of nebulae, clusters, and double stars and organizing them into a taxonomy as a naturalist would. And John Herschel certainly recognized that even though he had been published in their *Transactions*, few in the society understood his mathematical work. Baily lamented that in Britain there was "no association of scientific persons formed for the encouragement and improvement of Astronomy" and called for the formation of an astronomical society.[23]

Baily's appeal, outlining the merits a new society and the insufficiency of the Royal Society, was published in an appendix to his translation of a work by Italian astronomer Antonio Cagnoli (1743–1816), a context important for understanding the nature of what he and others felt Britain was lacking. Cagnoli outlined a method of using observations of lunar occultations to determine the shape of Earth more precisely. Not only was Cagnoli's work unknown in England prior to Baily's translation; it offered a straightforward project using simple observations to address an open question in terrestrial physics. Putting Cagnoli's method into effect would require coordinating observations among the growing number of observers in Britain—exactly the kind of project requiring vision and organization that the Royal Society was not providing. For Baily, the need was clear: here was a simple project with important potential dividends, and amateur observers in Britain had the means to contribute. They just needed a guiding hand.

Baily also lamented the state of the British *Nautical Almanac*, claiming it was inferior to those produced on the Continent and that, more damningly, the calculational tools it used were all imported: "*all* our astronomical Tables, *all* our Catalogues of stars, and *most* of our astronomical Formulae, in present use, are the production of the continent." This complaint resonated with Herschel, who had pointed out the same thing with respect to mathematics when forming the Analytical Society. Knowledge was wealth, and calculational efficiency was productivity—and in both, according to Baily's evaluation, British production was demonstrably inferior to that of the Continent. Neither the government, in the form of the Board of Longitude and the Greenwich Observatory, nor the scientific community, in the form of the Royal Society, was effectively utilizing the astronomical capital of the nation. For Baily, it was time to create an organization that would do just that.[24]

Baily was not the first to urge the formation of such a society. William Pearson (1767–1847), an amateur astronomer and schoolmaster, had floated the idea to other schoolmasters as early as 1812. A society of schoolmasters, as much as a society of stockbrokers, was outside both the gentlemanly sphere of the Royal Society and the institutional framework of Oxford or Cambridge. What was needed to bring this vision to life was someone with energy, enthusiasm, mathematical acumen, and a known desire for reform who had the credibility to bridge the divide between middle-class professionals like Pearson and Baily and the elite of British natural philosophy—in other words, someone like Herschel.[25]

A NEW SOCIETY

Along with Herschel's experimental work when sunlight was available to stream through his polarizing apparatus, by night the weight of his father's expectations hung over him to continue his astronomical endeavors. Despite his promise to take up William's observations, actually carrying this out remained a sidelight to Herschel's other pursuits. Yet as Baily and Pearson began to advocate for the creation of an astronomical society in London, Herschel became a central organizing figure in their movement. For Herschel, the new society proved the ideal synthesis between his desire to apply analysis to the physical world and his father's urgings to continue work at the telescope.

Herschel learned of Baily's ideas from his old friends Peacock and Babbage when the two came for a stay at Slough at New Year's 1820. Peacock, still smarting from negative reactions to their mathematical reforms at Cambridge, continued as a tutor and lecturer at Trinity College. He had recently taken clerical orders and was advocating for the establishment of a university observatory. The three friends would have visited with William Herschel, who no doubt again made the case for the importance of continuing his work, and they likely also discussed the governmental and institutional support for natural philosophy Babbage and Herschel had seen on their trip to France. It was easy for the two to convince Herschel of the necessity of Baily's new society and get him to commit to become a founding member.

On the evening of Wednesday, 12 January, Herschel and Babbage met with Pearson, Baily, and a handful of others for dinner at the Freemason's Tavern in London to discuss the details of the proposed society. Baily had assembled an eclectic mix of individuals. Their numbers included Henry Thomas Colebrooke (1765–1837), the son of a banker who had spent a portion of his career in India analyzing British colonial economic policy and publishing Sanskrit astronomical catalogs; Benjamin Gompertz (1779–1865), a mathematician and actuary who championed the use of negative and imaginary numbers in his calculations to the chagrin of more conservative calculators; and Patrick Kelly (1756–1842), a mathematician and friend of Herschel's father whose membership to the Royal Society had been blocked twice.[26]

For Herschel, the composition of the new society contrasted sharply with both the savants of the French Académie and the social elites of the Royal Society. He found himself among a group of London middle-class professionals motivated by meritocracy, private business ideology, and accountability, who planned to apply their forms of mathematical bookkeeping to astronomy. In terms of their practice and pursuits as well as their social status, most were marginal to the culture and endeavors of the Royal Society. Yet Herschel saw that this group of stockbrokers and businessmen had the resources and energy to employ his mathematical reforms in the business of astronomy. The ideals of the Analytical Society found a new guise and direction among what would become the Astronomical Society of London.

This dinner conclave was not the society's official birth. It was only a planning meeting, but the attendees left with clear directions for next steps. They would publicize their intentions with a declaration distributed throughout Britain inviting anyone interested in their aims to join the society. Perhaps recalling the ultimate failure of the Analytical Society, which had begun only informally and extended just to students at Cambridge, the new society wanted to solicit membership as widely as possible from the very beginning. Herschel was charged with composing this initial declaration of intent. Within days of the meeting, Herschel asked Babbage to send his astronomy papers to Pearson's, where he was staying in London. The next morning, when Babbage brought them, he and Herschel read over a draft Herschel had readied, perhaps recalling the similar prospectus they had prepared together for the *Memoirs* of their Analytical Society.[27]

In this new address, Herschel's theme was the need to coordinate and collaborate astronomical observations. As Herschel knew from his father's work, individual observations were of limited use if not part of a larger program. Telescopic observations were "useless capital" unless they could be confirmed and compared by other observers. As instruments and interest in astronomy spread and new observers began their own separate projects, Herschel knew the majority of them would be fruitless unless there was a means of collecting, coordinating, and standardizing their results. Even something as seemingly straightforward as positional star catalogs, for instance, were only effective if star positions were reduced and presented in a uniform fashion.

The new Astronomical Society, according to Herschel's vision, would provide this clearinghouse for data and coordinate observations of the entire heavens. This would keep observers from wasting time and effort repeating what had already been done. Efficiency of labor was a central virtue for these new business astronomers, and their society would ensure that labor was most effectively divided among observers. Coordinated astronomical surveys were essential, Herschel argued, because, in spite of his father's work, beyond "our own system, all at present is obscurity." In the same way collecting geological observations was necessary to construct theories of Earth's past, astronomical observations coordinated by the society were necessary to produce theories of "formation and motions of sidereal systems." William Herschel's independent, uncoordinated observations were not enough to do this, just as a geologist working alone without comparison and coordinating with others could not frame a theory of Earth.[28]

Imperial and commercial considerations were also at play. Besides new knowledge that would emerge from coordinated astronomical surveys, the society would diffuse methods of positional and calculational astronomy to British observers "scattered . . . over the surface of the globe." As British imperial and mercantile power continued to expand, an Astronomical Society based in London would connect and coordinate British observers around the world. Making successful observations of stellar phenomena like solar eclipses and lunar occultations required observers spread abroad far beyond Great Britain. The society also planned to coordinate communications with foreign astronomers and publish regular bulletins (something the *Nautical Almanac* failed to do) to alert observers at home and around the world to significant astronomical events. At their initial meeting, Baily had no doubt expressed his frustrations with the tables and data contained in the Board of Longitude's *Nautical Almanac*. The changes and improvements to it brought about by the new society would benefit not only astronomical science but the commercial interest of safe and successful navigation.[29]

The new society would be important for British industry as well. Besides center of a commercial empire, London was also home to a growing market for precision astronomical instruments. In the second half of the 1700s, British instrument makers had risen to the height of their fame, and any observatory in Britain or on the Continent contained telescopes or meridian instruments from their London workshops. By 1820 though, this dominance was threatened by Continental artisanship. The

Astronomical Society would give London instrument makers like Edward Troughton (1753–1835), who quickly became involved in the society, a dedicated users' group and a chance for its members to compare the relative merits of their instruments. For the astronomically minded businessmen, many of whom were pouring personal fortunes into the collection of expensive instruments, the society was a way to both ensure they were making informed purchases and to create a potential resale market.[30]

After the initial planning dinner, a follow-up meeting was held among the handful who would become the new society's leadership. Herschel was not present at this meeting, which was held Saturday morning, 24 January, but Babbage wrote him a summary of the decisions made. Baily and Babbage were both made secretaries, and Babbage told Herschel he had planned to nominate him for the position of foreign secretary and that the others had already had the same idea. The role of foreign secretary was essential for the new society to carry out the proposed coordination and standardization of observations of astronomers around the world. The position needed someone with a measure of international standing already. Herschel, on the strength of his family name, his mathematical accomplishments, and the contacts he had cultivated on his recent visit to France, was the obvious choice. As foreign secretary, Herschel would not only play a mediating role between the business astronomers on the one hand and the savants of Cambridge and London on the other; he would be the connection between the fledgling society and the astronomers of Europe.

Responses to the preliminary address and applications for membership poured in from across Britain; however, the first general meeting of the new society was postponed due to the death of the king. King George III, the longest-reigning English king since the fourteenth century and William Herschel's patron and the funder of his immense forty-foot reflecting telescope, had lived in seclusion in Windsor Castle since his final lapse into dementia in 1811. On his death, the nation was gripped with grief. The business astronomers of the Astronomical Society signaled their respect by delaying the formal inauguration of their new organization and recognizing the king (as well as William Herschel's legacy) by making the huge telescope he had financed, which still sat behind the Herschel home in Slough, their society's seal and symbol.

Herschel traveled back to London to attend the rescheduled meeting, where he stayed again with Pearson. The next day Herschel's parents joined him in the city. William attended the society's first meeting, where the elder Herschel was voted vice president and Herschel's position of foreign secretary was confirmed. Herschel and the other secretaries were introduced to the Duke of Somerset, Edward Adolphus St. Maur (1775–1855), who was voted the society's first president. Somerset represented the balance the astronomers hoped to strike in their new endeavor. Though the society was created specifically to rebuff the Royal Society's lack of rigor and lack of recognition for merit in the pursuit of positional astronomy, the astronomers recognized the risk of being marginalized without a well-connected leader to lend their effort credibility. Somerset, a gifted mathematician and aristocratic fellow of the Royal Society, seemed the ideal choice.

Unfortunately, Somerset and the astronomers had underestimated Banks's opposition to what he saw as a threat to the Royal Society's hegemony over natural philosophy. Herschel had tried to allay these concerns in his initial address on the society's goals and purpose: his final line pledged that the Astronomical Society would avoid "all interference with the objects and interests of established scientific bodies." Moreover, a learned society focused on a specific field of natural philosophy was not without precedent. The Geological Society, for instance, had been formed in 1807. Banks had been annoyed with that step, but he refused to tolerate this astronomical revolt. He insisted to Somerset that "the Royal Society will be *ruined*" if the Astronomical Society were allowed to survive. Somerset, as Banks's close friend, felt compelled to step down.[31]

Banks's opposition underscored Herschel's belief that the current leadership of the Royal Society was actively standing in the way of scientific progress. Baily, in a frustrated letter to Herschel, pointed out that the Royal Institution, which had been established to promote public understanding of science and where Herschel had viewed the chemical lectures that had so distracted him while studying law, had initially been met with similar opposition but now served an important role in the London scientific community. The Geological Society had faced opposition as well. It was absurd, as Banks seemed to be indicating by forcing Somerset to step down, that someone could not be the member of more than one scientific organization or that the physical sciences could not be pursued by multiple societies.[32]

Herschel agreed with Baily's evaluation. He was shocked and disappointed by what he called Banks's narrow-minded behavior as he searched for other potential candidates who might be of suitable social position and free from Banks's sway. Fortunately, Herschel did not have to look far. William Herschel, who was clearly well connected and respected by a wide range of both mathematical astronomers and aristocratic natural philosophers, was the obvious choice. William resisted, saying his age and inability to fulfill even the most "trifling" presidential duties made this impossible, but he also made it clear his age was the "*sole motive*" for not taking up the position; he would not withhold support for the Astronomical Society out of loyalty to Banks. The elder Herschel eventually agreed to be the society's first president with the understanding that it would be a purely symbolic role with *no* duties.[33]

As the work of the new society got underway, the role of foreign secretary quickly turned Herschel into the coordinator of astronomical data flowing in and out of Britain. Almost as soon as he was nominated, Herschel wrote to Baily requesting information about observatories on the Continent. He crafted letters to the directors of as many of these as possible, knowing they were who he needed to be in contact with if the Astronomical Society of London was to coordinate international projects. He wanted among the society's foreign members "as *many heads of observatories as possible*." His letter campaign worked, and by summer the ranks of the society were being swelled with astronomers and observatory directors from across Europe.

This new role rapidly increased Herschel's correspondence network. He established ties with astronomers in the German states such as Friedrich Bessel (1784–1846), director of the Königsberg Observatory; Carl Friedrich Gauss (1777–1855), a mathematician from Hannover; Johann Elert Bode (1747–1826), director of the Berlin Observatory; and Karl Ludwig Harding (1765–1834), astronomy professor at Göttingen. From France Herschel received replies from Alexis Bouvard (1767–1843) and Jean Baptiste Joseph Delambre (1749–1822), both of the Paris Observatory. Herschel's astronomical network embraced observers as far as Henrik Johan Walbeck (1793–1822) in Finland and Barnaba Oriani (1752–1852) in Milan. Through this network Herschel not only had the opportunity to coordinate goals and observing practices; he also shaped international perceptions of the new society and British astronomy. The Astronomical Society connected Herschel with mathematical practitioners abroad more than the Analytical Society ever had.[34]

Even as the activities of the society picked up, Herschel still remained largely at his home in Slough, outside London. Here supervision of carpenters and masons in the construction of a laboratory provided an excuse to miss Astronomical Society committee meetings, to the frustration of Babbage, who told his friend he *must* come to the city occasionally to mix socially with natural philosophers. Yet Herschel preferred the quiet of Slough, spending his days conducting his optical experiments or riding. He and William continued slowly refurbishing the twenty-foot telescope. Carpenters were occupied "mending its ribs and splicing new bones on its battered skeleton." He had yet to begin any formal observing program with the instrument, but he observed with occasional guests and told Babbage after an especially impressive viewing session of Saturn and the moon that he could "now believe anything of the effect of reflectors of great aperture." Unfortunately, the wooden framework of the telescope needed more work than either Herschel or his father realized. When it literally came to pieces as he was trying to position the tube on the meridian, Herschel decided the framework needed a total reconstruction. "I have got an Astronomical Mania on me at present," Herschel told Babbage, "which I suppose is more intense than it otherwise would be from the impossibility of gratifying."[35]

Perhaps due to the repairs needed for his own instrument and the continued draw of responsibilities in London, Herschel soon began an observing partnership in the city that would continue for several years, despite the stormy disposition of his partner. In the summer of 1820, Herschel began staying over on his London visits at the residence of James South (1785–1867), a retired physician who devoted his time to astronomy. South had a passion for instrumentation and was immensely proud of his two equatorial telescopes, with components crafted by London instrument makers Edward Troughton and Peter Dollond (1731–1820). Eager to put them to work, he and Herschel began a project of reobserving all of William Herschel's double stars, measuring the angles and distances between pair components. Unlike Herschel's own instrument, which was well suited for sweeping areas of the sky but not seeking out specific targets, equatorial telescopes were mounted so the telescope could track an object across the sky. This allowed for precision measurements of any double star in the field of view.[36]

Measuring double stars was exactly the sort of coordinated project the Astronomical Society had been created to cultivate. Double stars could be studied by a wide range of observers but had been neglected since the elder Herschel's early discoveries. They also offered physical insight into the sidereal universe: observations over time could determine whether the components of any double stars had moved, and mathematical analysis could show whether this motion was due to Earth's motion around the sun or the mutual gravitational attraction of the two stars. If it was the former, this could provide the first direct measurement of stellar parallax and be useful for calculating stellar distances; if the latter, this would confirm the extension of Newtonian gravity to the sidereal universe.

Herschel's observational work with South also gave him the opportunity to apply mathematical analysis to his instruments as well as the objects he viewed through them. He compared the merits of reflecting telescopes, the kind constructed by his father and used at Slough, with the more common refracting telescopes, which he used in his work with

South. Refracting telescopes depended on the manufacture of quality glass free of imperfections, but it also depended on the elimination of aberration, distortions in the image that arose from the shape of the lens and the different refractive indexes of colors of light. Overcoming aberration was a significant challenge for instrument makers, a challenge that they viewed in terms of craft practice: something to eliminate through extensive trial and error with their results becoming closely guarded secrets.

Herschel, on the other hand, felt solving aberration was an ideal application of mathematical analysis to problems in optics. Aberration could be addressed mathematically, and in early 1821 he wrote a paper exploring exactly that. In the paper, Herschel noted that mathematics had not yet been applied to the improvement of manufacturing refracting telescopes. The process of correcting aberration, Herschel argued, was too complicated for opticians and instrument makers. Viewing telescope manufacturing as a craft in which problems were solved by trial and error instead of by the application of theory was no longer sufficient. To address this, Herschel devised equations, emphasizing "symmetry and simplicity," that opticians could use to determine the shape of lenses to eliminate aberration. Instead of artisan methods, Herschel's equations reduced lens-making to a science for producing lenses of any specified focal length, radius, and index of refraction. Of course, he admitted, this depended on being able to manufacture glass free of flaws and with a high refractive power, something that was still a technical challenge, but Herschel hinted that government funding for research in glass manufacturing would soon make this a reality. Combination of mathematical analysis with government support would ensure the continued superiority of British instruments.[37]

Herschel soon realized the truth of Babbage's advice that he spend more time in the city. Despite his desultory astronomical work in Slough, the pace of his scientific and social life increased in London. He was soon observing with South for days at a time. In addition to astronomy and the organizational tasks of the Astronomical Society, Herschel also found himself at the center of an attempt to reform British science when the death of Joseph Banks in June of 1820 ended a forty-year presidency that Herschel and his colleagues felt had long outlasted its usefulness for scientific progress. Whoever took Banks's position as next Royal Society president would have the power, as Banks had, to shape British natural philosophy in their own image. If Herschel could influence the outcome of the next election, it would be a significant step for modernizing and reforming the society.

THE WOLLASTON COUP

Herschel and his colleagues wasted no time. The most likely candidate for the position of Royal Society president was Humphry Davy, who was, on the surface, the kind of candidate Herschel and his colleagues in the Astronomical Society would be expected to support. Davy came from a non-aristocratic background (though he was made a baronet in 1818 in honor of his services to science and industry) and gained fame as a chemist and the inventor of a lantern that could safely be used in coal mines, an invention that saved hundreds of lives and was essential for the British mining industry. Davy had lectured extensively at the Royal Institution, where Herschel had seen his large voltaic pile used to separate substances and isolate a whole series of elements for the first time. Here was a true scientific practitioner who could bring scientific rigor to the Royal Society.

Unfortunately, despite Davy's humble background and his achievements in natural philosophy, he also embodied many of the vices Herschel felt were hindering the Royal Society. Davy's personal character was, according to Herschel, "arrogant in the extreme." Even more damningly, he was not an impartial supporter of the sciences. Rather, he was "impatient of opposition" to his own views, and if put in power, Herschel was certain, would "oppose rising merit" that threatened his own line of research. Perhaps worst of all, Davy was fond of arguments and seemed a magnet for divisiveness. A Davy presidency would, Herschel feared, involve the Royal Society "in controversies of much personal acrimony with the other learned European bodies." For all Davy's chemical successes, he lacked the virtues that made for a true natural philosopher and as such needed be blocked from the presidency if at all possible.[38]

Herschel and his colleagues greatly preferred William Hyde Wollaston, whom Herschel had come to know while in London studying law. Wollaston was the son of a well-known amateur astronomer and, like Herschel, a Cambridge graduate. Like Davy, Wollaston was a chemist, and business astronomers like Baily were impressed with his financial acumen. (Banks, by contrast, had been shocked when Wollaston refused to disclose his method of platinum processing and instead profited from it.) Wollaston was already an active senior member of the Royal Society and, also like Davy, was from a non-aristocratic background. In contrast to Davy's arrogance and divisiveness, however, Herschel felt Wollaston would bring dignity, scientific accomplishment, and the reformers' values of efficiency and transparency to the presidency—assuming he could be convinced to stand for election.

The day after Banks's death, Herschel met with Peacock at Babbage's home in London to discuss what to do. The friends agreed Wollaston should stand for election against Davy, and Babbage and Herschel immediately called on him to convince him to do so. He was out, but they returned again that evening, found Wollaston at tea, and asked him outright whether he would allow his name to be put forward. The older man told them he had no intention, pointing to a scientific instrument and saying the presidency would interfere with "his little pursuits." When pressed, he admitted that if it was for the good of the society he might consider.

This was enough of an opening for the conspirators, who decided they would force Wollaston's hand by canvasing members of the Royal Society for Wollaston votes. When they had gathered enough to make it clear he was the preferred candidate, they would approach Wollaston again and ask him to stand for the sake of the society. Accordingly, Herschel began a drive to promote Wollaston's candidacy. With Peacock, he wrote notices for London papers, called on society members, and secured promises of support. At the end of their first day campaigning, Peacock set off for Cambridge, pledging to solicit support for Wollaston there, and Herschel stayed up until 2 a.m. the following morning along with Babbage and other colleagues writing nearly one hundred letters to Royal Society members urging them to support Wollaston.[39]

After two days of ceaseless campaigning, Herschel again called on Wollaston and shared what he and his colleagues had been up to, begging Wollaston at least to do nothing to act against them. The old chemist was still undecided, and Herschel emphasized to Wollaston the importance of keeping the society "from becoming an arena for all sorts of conflicting interests," which he felt it would become under Davy. Herschel also emphasized the support Wollaston could expect from Cambridge and the fellows of the Astronomical Society. He left the meeting encouraged to keep up his efforts.

Davy, however, was affronted by what he referred to as "junior members . . . disturbing the peace of the Society." In a personal remonstrance with Wollaston at a society dinner, Davy accused him of recruiting friends to campaign on his behalf. Such an obviously self-serving step was against the scruples of polite society: one's friends could bolster support for a particular candidate, but not at one's own request. Wollaston protested this was not the case, but he asked Babbage and Herschel to cease their efforts on his behalf. In an attempt to avoid a contested election that would be contrary to the gentlemanly conduct of the society, Wollaston suggested the matter be settled by chance or outside arbitration. When Davy demurred, Wollaston backed down. Wollaston was nominated to serve as president, but only in the interim before Davy's formal (and practically uncontested) election.

Wollaston justified his decision to withdraw by maintaining that taking on the responsibilities of Royal Society president would have been detrimental to his own scientific pursuits, his independence, and his personal happiness—all considerations that would come back to haunt Herschel when he eventually found himself in a similar position. In his attempt to influence the election, Herschel had realized the importance of connections among London's scientific practitioners, but he had also learned that these connections could be futile against the tempestuousness of personalities at the center of the society. This would come back to haunt him as well when Davy's impetuosity collided with Babbage's temper to create a rift that would nearly destroy Babbage's long friendship with Herschel and threaten the society's credibility in the eyes of the public.

For now though, Herschel had moved further into the London scientific orbit. Without Banks's opposition, many members of the Astronomical Society now also became members of the Royal Society. The astronomers' revolt against the Royal Society transitioned into a reforming movement *within* the society as Herschel was elected to the Royal Society Council. As a member of the council, Herschel had an opportunity to reform the practice of natural philosophy by reviewing papers read at meetings, recruiting reviewers for scientific articles, and helping determine whether submitted papers were published. This placed him in a position of leadership within both Royal and Astronomical Societies.

It had been an active year for Herschel and a profitable first year for the Astronomical Society. Baily's brainchild had grown to more than one hundred members, with Herschel ensuring their ranks included the foremost European astronomers and mathematicians. New observatories were being founded at Cambridge, bringing that bastion of conservative mathematics into the astronomical fold, as well as the far reaches of the British empire, at the Cape of Good Hope in South Africa, offering British observers the means of creating precision star catalogs for the southern skies. As the society's first annual report stated, "Until every remarkable star in the heavens is recorded, and its placed assigned . . . it is vain to pretend to an accurate knowledge of the true system of the universe." Much remained to be done. Yet besides Herschel's work in both societies and his observing program with South, there was an additional pull away from his father's telescopes at Slough to the social circles of the metropolis.[40]

As 1821 began, the scientific reformer had fallen in love.

CHAPTER 4

VALLEYS AND SUMMITS

They were no idle scamperers on the mountains that made these wild recesses first known. . . . It was a desire for knowledge that brought the first explorers here.

—JOHN TYNDALL, *FORMS OF WATER*, 1872

"Good God, Babbage," Herschel had fumed in 1814 in response to his friend's announcement of his sudden marriage, "how is it possible for a man to calmly set it down [in] two sentences, add a few more which look like self-justification, and pass off to functional equations?" Twenty-two-year-old Charles Babbage had fallen in love with Georgiana Whitmore (1793–1827) and, much against his father's desires, married. The elder Babbage, who supported his son financially and insisted he be economically self-sufficient before marriage, was furious. Though the marriage proved a happy one, Babbage's decision created a rift with his father not healed for years. For the dutiful Herschel, such an impetuous move seemed almost inconceivably reckless.[1]

By 1821, seven years later, many of Herschel's closest friends, including James Grahame and Richard Jones as well as Babbage, were happily married. Herschel, meanwhile, claimed to value the independence of his bachelorhood. He told Babbage he had fallen in love once and that it had cost him "pain enough . . . to eradicate one hastily conceived and silly passion." He would not let it happen again. If he were ever to become romantically involved, he assured his friend, it would only be "the result of long acquaintance . . . preceded by perfect and intimate friendship." Yet romance would in fact soon plunge Herschel into social and financial

96

turmoil. Herschel's engagement that year to Harriet Gwatkin, daughter of a wealthy landowner, placed him at the center of awkward negotiations that would prove too much for the natural philosopher. The engagement was broken off, to his disappointment and humiliation. Herschel's naïveté, his friends and family felt, left him adapted for analytical equations but largely helpless when it came to worldly concerns.

This personal disaster provided the excuse for a return to the Continent with a trip orchestrated by Babbage. This time, instead of the two savants focusing on Paris, they embarked on an extended Alpine expedition. Their journey, which saw Herschel go from the depths of personal despair to becoming the first Englishman to summit (so he believed) Europe's second-highest peak, was part of a tradition of British natural philosophers transforming the Alps into the laboratory of nature. Herschel used the mountains as a test case for his new approach to natural philosophy: capturing data; quantifying heights, distances, and physical properties; speculating on the structure and mechanics of glaciers; and documenting vistas using the camera lucida, a drawing-aid precursor to photography. On this trip, Herschel began to forge a Romantic view of Alpine physical sciences as an embodiment of his approach of bringing mathematical tools to understanding the world. The cold, clear heights of the Alps, hundreds of miles from the messy realities of metropolitan life, gave Herschel the chance to refine his ideas regarding natural philosophy and extend his network of astronomers into northern Italy and the Piedmont.

When Herschel returned from this long trip abroad, it would be to settle into an uneasy equilibrium in London in which he sought to balance several different aspects of his scientific activity. He turned down what he had sought multiple times in the past, a position at Cambridge, realizing he would have more opportunity to pursue natural philosophy independently. As a member of the Board of Longitude, he found himself navigating the increasingly fraught relationship between that body and the new astronomical practitioners of the Astronomical Society. As his astronomical practice increased, Herschel began functioning as a champion of reflecting telescopes while continuing to investigate the optics of refracting telescopes. He was awarded the Royal Society's prestigious Copley Medal for his mathematical work, which honored the same mathematical labors he had now come to see as too narrow for his view of natural philosophy. Even as Herschel achieved scientific prestige, he recognized the need for continued reform in the structure and institutes of scientific practice.

The catalyst that would ultimately direct Hershel's energy down a specific channel was the tragic death of William Herschel—tragic because Herschel was again abroad when his father died. William's death drove Herschel back to the telescope, where he finally began the project that was to consume much of his long career. Far from excluding other work, however, Herschel's resumption of his father's vast project to survey the heavens provided the framework by which Herschel could systematically reflect on the nature of light and ultimately the nature of science itself.

A BROKENHEARTED PHILOSOPHER

On 26 May 1821, Christie's auction house in London held a sale of the third and final portion of the estate of the late Mary O'Brien, Dowager Marchioness of Thomond (1750–1820). The marchioness was niece and heir of the British painter Sir Joshua Reynolds (1723–1792) and had inherited his drawings, paintings, and sketchbooks. At the auction, the lot for sale included more than 400 of Reynolds's drawings and 9 of his sketchbooks. Among these were many sketches of another Reynolds niece, Theophila Gwatkin (1757–1848), sister to Mary. Theophila had been a favorite of the painter and sat as model for many of his most well-known paintings. John Herschel, at the age of 29 still financially dependent on his parents, purchased nearly the entirety of the Reynolds lot for sale, spending the equivalent of over $22,000 for 300 drawings and 6 of the sketchbooks. For a bachelor with no independent income who had complained before about the cost of chemical equipment and supplies, this was a significant investment.[2]

The purchase had less to do with Herschel's artistic tastes than it did his interest in Harriet, one of the daughters of Theophila and her husband, Robert Lovell Gwatkin (1757–1843). Robert Gwatkin was a Cambridge graduate and a relative of Herschel's friend and fellow Analytical Society member Richard Gwatkin. Herschel likely met the Gwatkins in London through Richard and became acquainted with their seven daughters. By the spring, he was courting Harriet, the fifth Gwatkin daughter. Marriages in the early 1800s, however, were delicate arrangements, especially if they were potentially between the daughter of a wealthy Cornish landowner and reforming politician and the son of an immigrant astronomer. Though Herschel's mother's fortune made the arrangement more feasible (and allowed Herschel to purchase the Reynolds artwork), there were still fraught financial details to navigate. A marriage settlement was required to specify what funds were to be held in trust to support the bride and groom, as well as the situation that would entail upon the groom's death. These details were essential for safeguarding the bride, who would disappear as an independent legal entity during the marriage. Though in the case of Herschel and Harriet both were of legal age to marry, Harriet would have had no financial independence, and Herschel, though he stood to inherit a sizable fortune from his parents, was not yet financially independent himself. The financial situation for the couple would be determined by their fathers, and it was not uncommon for proposed marriages to fall apart when these arrangements could not be settled.

On 6 May, weeks before the Christie's auction, Robert Gwatkin called on the elderly William Herschel at Slough to discuss the marriage of his daughter with William's only son. William had been agitated that morning after receiving a letter from Herschel, in which the young philosopher may very well have surprised his parents with news of his engagement and Gwatkin's impending visit, but the old astronomer had calmed by the time Gwatkin arrived. "The meeting," Mary Herschel related, "went off better than expected." The conversation between the two was short, with both fathers agreeing the couple should be given more time "before anything should be talked over and settled." William, Mary went on, was "perfectly satisfied in your choice of the Lady and Mr. G. likewise spoke of you in the most handsome term." There seemed to be "no impediment in the way to prevent" the marriage.[3]

It was against this background that Herschel attended his first Christie's sale with the Gwatkins and returned two days later to purchase a self-portrait of Reynolds for 100 guineas (nearly $12,000). Two days later, on 20 May, he visited "Mr. G" to "make arrangements." These might have been financial arrangements related to the engagement, but it is also possible that Gwatkin wanted to discuss the next auction. The bulk of the Reynolds drawings would be on the block, and the Gwatkins no doubt wanted the materials to remain in the family without the potential awkwardness of having to bid on the property of Theophila's dead sister. If a gallant suitor were to make the purchases on the family's behalf (and subsequently marry into the family), this would circumvent the awkwardness. Unfortunately for the Gwatkins' hopes, though Herschel made the purchases, the sketchbooks never made it back into their family.[4]

The problem appears to be that though Herschel apparently had no issue spending money in his pursuit of Harriet, he had immense difficulty talking about it. Though he would later become known for his financial astuteness and go on to become master of the mint, at this point in his life Herschel had far less sense regarding finances than his friends and family hoped and apparently not enough to navigate the rocky shoals of a marriage settlement. Two months after the meeting between his father and Robert Gwatkin, Herschel asked Babbage to become trustee for the marriage settlement between him and Harriet. Trustees were custodians of the financial resources set aside by the bride's and groom's families to support the couple in their marriage, charged with overseeing the invested funds that would support the new couple. Herschel had met with his future father-in-law and did not think there would be any issues. "The arrangement is of the simplest kind," he assured Babbage, "and in all probability will never be productive of trouble to any of the parties concerned in it." Despite the fact that Babbage had married young and against his own father's wishes, Herschel trusted him and knew he had the financial acumen to serve in this role well.[5]

Herschel was especially appreciative of Babbage's discretion and diplomacy, which had recently saved Herschel from becoming embroiled in an embarrassing feud between his colleague James South and the instrument maker George Dollond (1774–1852). After an argument had broken out between South, whose explosive temper would eventually alienate him from Herschel, and Dollond at a meeting of the Astronomical Society, Herschel felt obliged to become involved on South's behalf. In a long letter, Babbage took Herschel to task for his response, telling Herschel he was acting "with a degree of harshness quite foreign to your character and which I am sure at some further period you will sincerely regret." On the letter from Babbage, which he preserved, Herschel wrote: "This is the only letter relative to the *abominable* business it refers to I have preserved. It is the letter of a true friend." This incident was one of the few times that Babbage, who was usually of the two the more impulsive and irascible, rebuked his more reserved friend for bad behavior. The reproach only increased Herschel's measure of Babbage's judgment, something he would rely on his friend for in the negotiations with the Gwatkins.[6]

Because Herschel's father was unwell and some distance from London, Charles and Georgiana Babbage fulfilled an important social function in the negotiations as well, hosting the Gwatkins in lieu of Herschel's parents. As things progressed, Babbage warned Herschel that the terms of the marriage settlement needed to be very clear. Since, as he said, Herschel never paid much attention to money, Babbage spelled out the details exactly, explaining what kind of settlement he would set up if he were in Herschel's shoes. In particular, Babbage wanted a settlement that gave trustees active power over how the money was invested. Getting these details right was essential. But Babbage's instructions forced Herschel, who by Babbage's own admission had little head for finances, into the awkward role of negotiating the details of the settlement with his future father-in-law. Babbage urged Herschel to lay aside his customary reserve and discuss the matter openly with Gwatkin.[7]

Babbage was not the only one worried about the financial details of the impending marriage. A week after Herschel asked Babbage to act as trustee, Babbage received a note from Herschel's mother asking him to ensure the bride's fortune was "properly placed in funds for their use." She added that Herschel was "so perfectly unacquainted with business of this kind, that it will be more necessary for you to be so good as to see this business properly executed." She also asked Babbage not to share her letter with Herschel, who would have been embarrassed to find that both his parents and closest friend felt he was not up to negotiating a marriage settlement if left to his own devices.[8]

Despite the worries of friends and family, Herschel went to work in London making the requisite financial arrangements. A detailed letter from his mother provided instructions on how to gain access to funds from an inheritance from his great-aunt, his father's money at the Treasury (likely a portion of his royal stipend as the king's astronomer), and his parents' money at the bank. Having collected from these various sources, Mary Herschel hoped "all money matters will be finally settled between you and Mr. Gwatkin." Babbage was unsatisfied, complaining to Herschel's mother that Herschel had still not provided the details of the settlement and that Babbage had "felt obliged to make some remarks to him which could not fail to be unpleasant." Babbage's frustration only mounted as Herschel's reserve and lack of communication ultimately stymied negotiations.[9]

Herschel was unable to bring the clarity and precision he and Babbage had pursued in their mathematical reform to his human and financial relationships. Babbage's assistance was apparently not enough to untangle whatever difficulties had arisen, and in July 1821, months after the purchase of the Reynolds drawings but only days after the flurry of letters between Babbage and Herschel's mother, the engagement was abruptly broken off. According to a later account, Herschel shut himself up in a dark room for days and would see no one. The Reynolds drawings, the purchase of which was significant enough to have been mentioned in the London papers, were never passed on to the Gwatkins and remained in Herschel's possession, no doubt a painful reminder of the emotional perils of navigating social expectations among London elites.[10]

AN ALPINE ESCAPE

Whether it was Herschel's reticence or Babbage's financial advice or both that contributed to the broken engagement, Babbage took responsibility for comforting his friend and extricating him for a time from London society. The solution to Herschel's embarrassing heartbreak was to travel once again to the Continent, this time in the company of a few close friends and with the intention of going beyond Paris to pursue measurement and experiment in the Alps. Whereas their previous trip had been one of scientific diplomacy, meeting the leading scientific lights and conversing on optics and analysis, this second trip with Babbage would take them into the mountains to geologize and apply their tools for precision measurement and calculation to one of the most important heavenly objects: Earth itself.

Herschel and Babbage had already developed an interest in geology, which was becoming an important field of study during the early 1800s. Though Britain remained somewhat isolated in terms of developments, the Geological Society of London had been founded more than a decade earlier, and there was a growing interest in understanding both the chemical properties of mineralogical specimens (which Herschel had pursued at Cambridge) and the processes that led to their formation. Mines, canals, and railway cuts across the countryside exposed engineers and natural philosophers to the distinct layers of Earth's surface, and in 1806 the English surveyor William Smith (1769–1839) had produced the first detailed geological map of all England and Wales, showing that it was possible to connect common rock formations and fossils across larger regions than ever before. In 1820 Babbage and Herschel had planned their own geologizing trip through southern England, with Herschel confessing at the time that he was "as innocent as the unbegotten babe of any tincture of geology." His ignorance, he joked to Babbage, would not be an issue, however, as the field was so new that "one may now go blundering on in security, satisfied that nobody knows what rocks *ought* to be found formed in such & such situation."[11]

Geological investigations in England were characterized by a reticence to theorize and a tendency to interpret new geological knowledge in light of traditional Biblical claims of a worldwide flood. When the Geological Society began publishing its *Transactions*, it emphasized the collection of "facts" needed to test theoretical "opinions." Many English savants pursuing geology resisted posing overarching geophysical theories at all, content to gather precise data on fossils and rock formations. At the same time, Oxford was becoming a center of the teaching of geology, as professors such as William Buckland (1784–1856) and William Conybeare (1787–1857) began doing (and teaching) fieldwork. Oxford was also a traditional center of Anglican theology, so for early geologists like Buckland and Conybeare (who were both clergymen), evidence of cataclysmic events in Earth's geological history were interpreted if not as direct proof of Noah's flood at least as showing strong evidence of the effect of tidal waves or immense flooding events. Phenomena like giant boulders that clearly had been carried miles from their source rock or valleys that could not be explained by the slow and steady erosion of the streams within them were seen as evidence in this direction, and these were the kind of sights Herschel and Babbage were eager to see on their trip.[12]

Their earlier geological expedition never materialized, but Herschel's broken engagement gave Babbage the perfect excuse to plan a much more extensive trip during which they could explore geological formations not found in Britain and dip their toes into some of these questions about Earth history. The two would return to Paris and then travel onward through the Alps and into Italy. Herschel and Babbage left London on 21 July. This time they were joined by their friends George Peacock, still continuing the Analytical Society's mathematical reforms at Cambridge and who planned to travel with them as far as Paris, and Richard Jones, still living the quiet life of a country parson but beginning to write on political economy. Unfortunately, Jones had no passport and so accompanied the three friends for the Channel crossing alone. The party planned to sail by steamboat from Brighton to Dieppe, on the northern coast of France, but at the last minute changed their plans. The decision turned out to be fortuitous, as the steamboat caught fire and burned the day they would have been on it. The four travelers avoided disaster but nonetheless had a miserable crossing, reaching Dieppe in the early hours of the morning "after a passage of thirty-one hours in which everything . . . which could render a sea voyage horrible was united." Everyone on board, even the sailors, was sick.[13]

Two years earlier, Herschel had arrived in Paris as an enthusiastic mathematician with an interest in optical experimentation. Now, he had official roles in the Astronomical Society and the Royal Society, along with official business to conduct. In his position as council member for the Royal Society, Herschel had been charged with coordinating a joint project between the Royal Society and the Bureau des Longitudes. The goal was to triangulate the exact relative geographic positions of Greenwich Observatory and Paris Observatory, in effect linking together surveys of the two countries. In this role Herschel unknowingly stepped into a feud that had erupted since his last time in France between the two astronomers he had assumed would have the largest roles in the project. François Arago and Jean-Baptiste Biot, who had once traveled together conducting similar triangulation surveys, had since fallen out over results in the field of optics. Their conflict illustrated the first cracks in the Laplacian goal of unifying all physics, but it also involved rivalries about priority and the rights to pursue certain research areas that Herschel felt were unworthy of natural philosophy and those who pursued it.

Biot had returned from a trip abroad to find Arago conducting research in an area of optics Biot considered his own territory, and Biot fiercely defended his claims of discovery against his French colleague. When Herschel arrived, Arago pledged to participate in the surveying project, but Biot explained it would be impossible for him to help as long as Arago was involved. Herschel was eager to pick up where he had left off on his previous trip, discussing work on polarization with Biot, and was frustrated by the feud. Arago, in turn, vented his frustration to Herschel regarding William Pearson, one of the Astronomical Society's founders, who Arago felt was doing work on polarization without giving Arago sufficient credit. Though Herschel was happy to catch up on the latest experimental results, this trip to Paris gave him a sense of the bitter rivalries developing in optics and a strong distaste for the belief that research problems or topics could be considered personal territory. His dismay over colleagues claiming certain areas of study as their exclusive domain would soon work its way into his own writings on optics and his views on the practice of natural philosophy.[14]

Arago and Biot were also at odds regarding the nature of light itself. Biot's earlier work, which had inspired Herschel on his last visit to Paris, had explained the optical phenomena of polarization and double refraction in terms of Laplace's paradigm of intermolecular forces acting on light corpuscles. However, in the time since, Augustin-Jean Fresnel (1788–1827), a French engineer and mathematician, had explained diffraction, which had remained untreated by Biot's theory, in terms of the wave nature of light. Arago supported Fresnel's theory, which was ultimately awarded an Académie prize, against Biot. More than a controversy over the nature of light, Fresnel's explanation and Arago's defection was the first sign of the unraveling of Laplace's program for physics.[15]

At the end of their week in Paris, Herschel returned to Arcueil and dined once more with Laplace. Again William Herschel's work was the topic of discussion, and Herschel no doubt explained to Laplace and the other French savants that he had resumed his father's work with the twenty-foot reflector. This led Laplace to pronounce a verdict on the elder Herschel's work regarding double stars and the nature of nebulae as "truly philosophical and very true." If Herschel still sought confirmation that resuming his father's work was important to the wider scientific community, he received it that night from the reigning sovereign of French natural philosophers. (Babbage, on the other hand, was still having issues communicating in French. When Laplace spoke of Woodhouse, one of Herschel's mathematical influences at Cambridge, Babbage took him to be saying *vous deux*, referring to double star orbits, to the party's amusement.)[16]

Before leaving Paris, Herschel and Babbage received a distinguished visitor they had missed on their previous trip. The Prussian explorer Alexander von Humboldt (1769–1859), who had adopted Paris as a second home, had become since his travels two decades before one of the most distinguished naturalists on the Continent. His extensive journeys throughout Central and South America from 1799 to 1804 included studies of volcanoes and combined natural history, meteorology, and geology to bring about a new conception of the natural world. The naturalist, as he outlined in his 1805 work *Essai sur la géographie des plantes*, the first volume of results from his American voyage, should be a global physicist as well. It was not enough, Humboldt argued, to simply record data about specific places, species, minerals, or conditions. Humboldt sought to correlate worldwide patterns. Measurements of tides and temperatures, currents and weather patterns, distribution of species and heights of mountains, all became part of viewing the planet as a unified physical system—a view that had a direct influence on Herschel.[17]

Back in England, geologists were attempting to put together a history of the landscape of the British Isles, but Humboldt's travels were creating a conceptual map of the entire physical world. Herschel listened attentively to his accounts of scaling volcanoes in the Andes, no doubt hearing echoes of his old mineralogy teacher, Clarke, who had ventured up the slopes of Vesuvius collecting mineralogical specimens. Humboldt had wanted to do the same, but the Napoleonic wars forced him to go to an entirely different hemisphere to find volcanoes to study. Humboldt's bold vision of the natural sciences, and his conviction that volcanoes were nature's laboratories for understanding geological processes, would linger in Herschel's mind throughout and long after his Alpine expedition.

Babbage and Herschel left Peacock behind in Paris to continue on toward the true goal of their journey: applying their approach of measurement and observation to the grandest mountains in Europe. They hoped for more than simply impressive vistas (though they would certainly get those) or even the collection of prime geological or mineralogical specimens. Rather, they wanted to take the physical measure of Alpine peaks with every instrument at their disposal. From England, they had hauled with them a portable laboratory that included a mountain barometer, chest of chemical equipment, blowpipe for analyzing minerals, pocket compass and sextant for measuring heights, thermometers, mineralogical hammers, reflecting goniometer to measure the angles of the crystal faces of minerals collected, and a camera lucida with drawing equipment.[18]

The last of these instruments, the camera lucida, provided a compelling record of their travels and would become a tool Herschel used to document further travels for the rest of his long life. The camera lucida was an optical drawing tool in which an angled piece of glass superimposed an image of the observed scene onto a sheet of drawing paper. The artist (and it did indeed take a skilled artist) could then, by looking down through the glass with one eye and at the paper with the other, sketch a scene with a level of detail approaching the photographic. Before the development of photography (which Herschel also had a significant hand in), the camera lucida provided visual record of the landscapes through which they passed. Herschel was quite skilled with the tool, having been practicing with it since at least 1816, and produced dozens of careful drawings from this trip.[19]

From Paris, Herschel and Babbage traveled south by carriage. The summer heat, Herschel complained, was so great they were nearly "broiled alive." They stayed in Dijon for two days, and by their arrival Herschel was so fed up with their mode of transport that he decided to purchase his own carriage for the rest of the journey. He settled on a luxury carriage known as a landau with facing seats and a top that could be let down when the weather was fine. The cost was forty pounds, and Herschel hoped to sell it when their trip was concluded. From Dijon, their road led through the Jura Mountains, "clothed in forests of pines from the bases to their summits . . . broken into the most picturesque forms." Unfortunately, poor weather kept them confined to the carriage; otherwise, they would likely have spent time studying the limestone formations that later were used to link this region with a specific geological period—the Jurassic. The road wound "along the edges of precipices overhanging the most beautiful valleys" before finally descending toward Lake Geneva. Despite the poor weather, Herschel felt the view of the lake and surrounding vale worth the entire trip.[20]

The two travelers passed a week in Geneva, where they found "a great deal of scientific ardour afloat . . . and a very remarkable spirit of inquiry and improvement alive." Geneva, like many European cities, had formed a philosophical society, which met fortnightly, and Herschel and Babbage were invited to mingle with Geneva's leading natural philosophers. They mixed with Geneva's high society as well, being introduced to, among others, the historian Jean Charles Léonard de Sismondi (1773–1842), the British politician Gilbert Elliot-Murray-Kynynmound, Second Earl of Minto (1782–1859), and some Italian nobles who "had taken rather too active a part in the Neapolitan revolution & are now obliged to be anywhere but in Naples." Though everyone offered the travelers advice about mountain passes and ascents, the weather remained uncooperative, and after a week they had still not even glimpsed the imposing face of Mont Blanc, their first Alpine destination.[21]

In Geneva, and with the prospect of Mont Blanc before them, they would also have discussed the work of the Genevan explorer and naturalist Horace Bénédict de Saussure (1740–1799), the leading natural historian of the Alps. Saussure had been the first natural philosopher to summit Mont Blanc, in August of 1787, but more than that: he was the first naturalist to insist on the necessity of outdoor fieldwork for understanding geological processes. "Only the naturalist's own *firsthand* observation," so Saussure's reasoning went, "was adequate to comprehend large-scale features of the earth." Saussure had used barometric measurements to confirm Mont Blanc as the highest point in Europe and from his fieldwork came to believe that the folding of rock strata observed in mountainous regions testified to huge movements of Earth's crust in the past.[22]

Just when Herschel was beginning to despair that their expedition would be foiled by the weather, it finally cleared, and they could at last travel into the mountains. Herschel and Babbage quickly set off southward toward the valley of Chamonix. From here, Mont Blanc dominated the southern horizon, with the immense Mer de Glace glacier stretching down its flank and winding toward the valley's end. The golden age of Alpine mountaineering had not yet arrived and tourism to the area only existed on a tiny scale; Chamonix had at this time been visited by only a handful of English travelers hoping to climb the mountain and explore the glacier.

In addition to the massive bulk of Mont Blanc, which at almost sixteen thousand feet made it the highest mountain in the Alps, Chamonix was best known for its glaciers. The most spectacular of these was the Mer de Glace, today known as Montanvert, which at the time of Herschel's visit stretched over four miles down the northern side of Mont Blanc. At the beginning of the nineteenth century, glaciers remained a physical mystery, with their formation and motion not well understood. Herschel and Babbage, like Saussure, would combine careful physical measurement with their attempt to scale the peaks. Many would follow them, with later British physicists like James David Forbes (1809–1868) and John Tyndall (1820–1893) making Chamonix central to their theories of glacier flow. Within decades, the approach that Herschel and Babbage were pioneering on the Alpine slopes would become a central feature in British physics.[23]

Once they had gotten settled in the village, Herschel and Babbage set out with their guide—named Joseph-Marie Couttet, whose father had accompanied Saussure on his ascent of Mont Blanc—to explore the Mer de Glace, stopping frequently to perform a battery of physical measurements that included humidity, temperature, and pressure (from

FIGURE 4.1: Sir John Frederick William Herschel, *Aiguille de Dieux, Mont Blanc*. Graphite drawing made with the aid of a camera lucida, unframed: 7 3/4 × 11 5/8 in. The J. Paul Getty Museum, Los Angeles, Gift of the Graham and Susan Nash Collection.

which they could calculate elevation) with the time and location of each meticulously recorded. Herschel was overwhelmed by both the view—of perpendicular peaks of granite rising up on all sides—and the puzzle that glacier flow provided, and he and Babbage speculated about potential mechanisms as they hiked.

That evening they camped on the mountain with their guide in a hut overlooking the glacier. Babbage fell asleep almost immediately, but Herschel remained restless. After trying for a time to sleep, he finally gave up and decided on a foolhardy night ramble alone, climbing the slope above the glacier for a view in the moonlight. He described what he saw in the journal he kept during the first portion of the trip:

> The scene was the finest imaginable. The whole extent of the glacier in
> all its long windings . . . lay at my feet, gleaming in a moonlight brighter
> than I ever beheld. The long-receding amphitheatres of snow surmount-
> ed by granite pinnacles . . . and the dark contrast of the mountains
> opposite involved in shadow and standing off from the bright sky formed
> a combination it is hardly possible to describe. The stars shone with

FIGURE 4.2: Sir John Frederick William Herschel, *Vale of Suza, being the entrance into Piedmont descending from Mont Cenis*. Graphite drawing made with the aid of a camera lucida, unframed: 7 11/16 × 11 7/8 in. The J. Paul Getty Museum, Los Angeles, Gift of the Graham and Susan Nash Collection.

> a steady, untwinkling light, and Jupiter and Saturn were only distin-
> guished from the rest by their superior light. . . . The deep and lonely
> silence was only broken at intervals by the roar of avalanches echoing
> like distant thunder from below, and once, a single sheep bell.

It no doubt felt a million miles away from London and the scene of his heartbreak. After taking in the scene and the silence, Herschel start- ed back to camp but found himself unable to retrace his steps. He was forced to make detours around crevasses or impassible slopes that took him farther and farther from his path. Finally, realizing he was lost and that it would be dangerous to continue blundering about in the dark, he consigned himself to spending the night on the glacier and waited until morning to find his way back.[24]

Returning to the village of Chamonix, Babbage and Herschel lin- gered until it became clear that an ascent of Mont Blanc would not be feasible that season, at which point they continued southwest, through Chambéry and Aix-les-Bains, and over the pass of Mont Cenis, crossing from Switzerland into northern Italy. Along the way, Herschel was impressed by the engineering feat of the road itself. The ascent, beginning at the tiny town of Lanslebourg (now Lanslebourg-Mont-Cenis), rose almost to the snowline "by a series of inclined planes like a staircase cut in the hillside one above the other," broad enough to allow for the passage of artillery. Once over the pass, it fell again "along the side of a valley so rugged and abrupt one would have supposed it impracticable ever to have passed even on foot." Within three hours, Herschel reported with wonder, they had reached Susa, the first town of the Italian Piedmont. "The windings of the road," he recounted in a letter to his parents, "at last become so frequent and complicated as to be almost like playing at hide and seek." Like the night on the glacier, the scene captured Herschel's imagination: "Nothing can be more striking than this descent . . . which leads, between the rugged mountains of Savoy on one side, and a beautiful range of hills, all clothed with trees and crowned each with its ruined castle or picturesque monastery, on the other, into a boundless plain which spreads like a sea on the horizon."[25]

They reached the city of Turin, where Herschel and Babbage stayed two days, just long enough to repair Herschel's barometer that had been damaged at Chamonix. From there they continued eastward to Milan, where Herschel found the heat oppressive but the city magnificent. Here he turned his camera lucida from geology to architecture, using it to draw the city's cathedral, which he thought the best example of Gothic architecture he had ever seen. Having had a taste of mountaineering in Chamonix, the travelers also made more extensive preparations for a return north and further climbing, purchasing "cords, proper boots with hooked irons to screw into their toes . . . a long pole of 6 feet, pointed with iron, and an axe to cut steps in the ice."[26]

After several days in the city, they headed north again, traveling to Lake Como and arriving in a thunderstorm that made it impossible to see the scenery but thankful for their enclosed carriage. Here they met with Pietro Configliachi (1777–1844), professor of physics at the University of Pavia, and delivered correspondence from the Royal Society. Herschel had hoped to see more astronomers in Italy, but unfortunately Giovanni Plana (1781–1864), director of the Turin Observatory, and Francesco Carlini (1783–1862), director of the Brera Observatory in Milan, were traveling together in Austria performing measurements to link northern Italian maps with those of France, much as Herschel had proposed in Paris for the British and French.

Language continued to be an issue. Neither Herschel nor Babbage spoke Italian, so they got along as best they could with French, which sometimes caused problems: "I asked this morning for coffee, and as the waiter professed a perfect knowledge of French, I thought it not unreasonable to hope so mendicant a demand might be gratified. After feasting our imagination half an hour with the pleasing anticipation of the refreshing beverage, behold three men enter, one bearing a bundle of sticks, one a shovel & tongs, and one a quantity of lighted coals, to make a fire for the Signori Inglesi, in a temperature of 72." The weather finally cleared again, and they crossed Lake Maggiore by ferry with their coach under brilliant blue skies and in the midst of a landscape, Herschel described to his parents, such as "no one who has not been out of England can form a conception of."[27]

Passing back into Switzerland, the travelers headed westward down the Rhône Valley. At the town of Visp, they turned again south and began ascending the valley bounded by the massive Weisshorn on the west, the Matterhorn to the south, and their ultimate target, Monte Rosa, to the southeast. In Chamonix Herschel and Babbage had instructed their guide Couttet to meet them upon their return from Italy. Couttet said he knew a route up Monte Rosa, and with his assistance Herschel and Babbage were prepared to be the first Englishmen to make it to the summit, which was thought at the time to be as high as Mont Blanc and thus possibly Europe's tallest peak. Saussure had attempted it, though Herschel believed it had been summitted only once, by a French climber from Lyon a decade before. The latest attempt had been a German climber who had obtained only an inferior summit. This spoke to the difficulty of the climb but also the difficulty of distinguishing one peak from another or the highest summit in the midst of the surrounding prominences—a problem that would affect Herschel's party as well. (In reality, the highest peak of Monte Rosa, known today as Dufourspitze, remained unclimbed until 1855.) In Visp, they found Couttet waiting for them and set off up the valley of Saint Nicolai, with Monte Rosa looming in the distance ahead.[28]

After passing the remains of the village of Randa, destroyed by an avalanche a few years earlier, Herschel, Babbage, and Couttet reached the tiny hamlet of Zermatt at the valley's end. The next morning, they were ready to begin their ascent of the mountain itself. Herschel admitted he had little hope of actually achieving the summit but felt it worth the attempt. In Zermatt they hired four additional guides, rented mules, and purchased wine and provisions for a two days' journey. Setting off, they first traveled south over the ridge to the east of the Matterhorn, passing back into Italy at ten thousand feet and camping that night in a barn in the Italian village of Breuil-Cervinia. The plan was to resume the ridge in the morning and climb the escarpment to the northeast. The next morning, 7 September, they rose at dawn, their party increased by an old man hired in Breuil, and by ten o'clock had regained the ridge they crossed the day before. They took their breakfast on the ridge, with the expanse of the Piedmont stretching away to their south and the Matterhorn looming to the west, before beginning the climb of what they believed to be Monte Rosa itself. "We traversed waste after waste of snow," Herschel recounted, "without interruption and without limit, always mounting." After five hours of trekking though knee-deep snow, they reached a vantage from which they could stare down into the valley of Saint Nicolai from which they had come the day before. "The highest crest of the mountain now only remained to climb," Herschel recalled. "It was tremendous, being literally a sharp edge of snow along which a cat might have walked."[29]

At this point, the guides from Zermatt refused to go any farther. Babbage was exhausted as well, and Herschel convinced his friend not to push on. Herschel, however, was not ready to quit. He tied himself to the climber from Breuil, and Couttet scouted ahead, returning to report that continuing up the ridge they were on was too dangerous. With Couttet carrying Herschel's precious barometer, Herschel and the remaining mountaineers made a wide circuit and climbed for another forty-five minutes to reach a peak they believed was Rosa's summit. "Thus," Herschel reported, "I had the pleasure of being the only Englishman, and but the second or third individual, who has ever set foot on the summit of this stupendous mountain." It was not enough for Herschel to have simply made it up the mountain, though. More importantly, he (or rather, Couttet) had brought instruments to the summit as well, so Herschel was able to take barometer readings indicating the mountain's height to be fourteen thousand feet. This meant, perhaps disappointingly, that Herschel's peak was indeed second to Mont Blanc. But it also meant its altitude had been measured for the first time.

For Herschel, again as on the Mer de Glace, the scientific accomplishment was conjoined with Romantic vision: "Nothing the imagination can conceive can give the faintest idea of the scene. On one side all was clear and the eye lost itself in tracing mountains on mountain, all capped with snow, and valleys running into one another, each with its river like a fine silver thread down the middle. On the other, a sea of white clouds covered the plains of Lombardy and Piedmont, but rolling far below our feet, and only the highest snowy peaks of Mont Blanc and the loftier Alps emerging like islands among them."[30] It was an emotional and physical triumph. In two short months Herschel had gone from the despair and embarrassment of a broken engagement in London to standing at the summit of Europe.

The party returned off the mountain safely, though the time it took them to arrive back at Zermatt seemed much shorter than would be expected if they had indeed climbed Monte Rosa. Instead, it soon became apparent that Herschel and Couttet had instead reached the summit of the Breithorn, a peak to the west of Monte Rosa. The first ascent of this peak had been made only eight years before, by a group that included Couttet's father. Herschel realized this mistake fairly quickly, writing to the Earl of Minto, who he had met at Geneva, that he was "extremely scrupulous in asserting that I have ever even set foot on Mount Rosa" even though four guides and the old man from Breuil attested to it. The barometric results, in which Herschel had more confidence than the identity of the mountain he was on, seemed of insufficient elevation, though he had been certain there were no higher peaks in the vicinity. He scratched out the caption "Monte Rosa" and replaced it with "Breit-horn" on the camera lucida sketch he made.[31]

The next day their faces were sunburned and blistered, but Herschel was now convinced that their elevation and geological situation made mountains and glaciers the prefect natural laboratories. "Should I ever visit that part of Switzerland again," Herschel determined, "I shall oc-cupy Saussere's hut for some nights or pitch a tent on the vast plateau of snow on the s[outh] side of the highest ridge as it affords an admirable station for a multitude of interesting experiments." But for now, it was time to continue north. Herschel and Babbage left their guides behind and journeyed onward to Bern, from there making expeditions eastward to the lakes of Thun and Brienz and up the narrow valley of Lauterbrun-nen. They measured the height of the celebrated Staubbach Falls, where a creek spills over the lip of the valley so high it is lost as mist by the time it reaches the valley's floor. Taking multiple barometer measures, Herschel and Babbage meticulously measured this cascade from several different levels, eventually publishing their results. According to their measurements, the falls plummeted almost exactly one thousand feet in a single drop. There were similar measurements to be made at a second waterfall, that of the Giessbach on the Brienz, and a tour of the agri-cultural school of Philipp Emanuel von Fellenberg (1771–1844) outside Bern before the travelers headed home by way of France, complaining about the sluggishness of the postilions hired to drive their carriage.[32]

By the beginning of October, Babbage and Herschel were back in Paris. Their Alpine expedition had lasted almost six weeks. They had crossed the spine of the mountains four times and reached elevations achieved by only a few before them. Along the way, they captured the landscape through their instruments and Herschel's camera lucida sketches. Herschel's drawings and measurements allowed him to trans-form each experience into physical data. He was coming to realize that whether peering through a telescope at a new double star or scaling an

Alpine peak, the experience was incomplete if it did not result in standardized data. Even before arriving home, he was speculating about the possibility of skeleton forms or worksheets that could be distributed to climbers, allowing them to standardize the data they recorded in much the same way he would later devise similar forms for astronomers. His first Alpine trip sharpened his observing skills and instilled in Herschel a growing interest in geology as well as a conviction of the importance of physical experience married to quantifiable data. It also continued to shape his identity: when Herschel arrived back at Slough, he could add "seasoned Alpine explorer" to his growing reputation as astronomer and mathematician.[33]

AN UNEASY EQUILIBRIUM

Upon his return to England, Herschel resumed his responsibilities with the Royal Society. One of these roles was making comments on papers submitted to the society. There was as yet no formal system of review for papers published in the *Transactions*, but expert opinion from the council was a step toward the modern peer review process. Herschel was also busy with the Astronomical Society as it worked to fulfill one of the functions outlined in his initial address: a clearinghouse of astronomical information from around the world. In his role as foreign secretary, Herschel received communications of data and observations from astronomers outside Britain that were often included in personal letters to him rather than public communications to the society. Herschel argued that the extracts from these letters were appropriate to print in the society's planned memoirs alongside more formal papers, which they ultimately were. Thus in his role in both societies, Herschel helped shape forms of publishing that would in turn shape the scientific community.

The month after his return from the Continent with Babbage, Herschel was awarded the Royal Society's highest honor, the Copley Medal, for his "mathematical & optical papers" published in the *Transactions*. The Copley Medal was the society's oldest and most distinguished award, having been given annually since its inception in 1731. For Herschel, the award was a strong validation of the program of mathematical reform he had begun during his student days at Cambridge as well as his standing as a recognized mathematical authority in the London community of natural philosophers.[34]

Despite Herschel's growing scientific stature, he had still been unable to secure a position that would allow him to pursue mathematics as a full-time vocation. In 1820, before his broken engagement and subsequent travels with Babbage, the death of the mathematician Isaac Milner (1755–1820) had opened up the Lucasian Chair of Mathematics, one of the few livelihoods for a mathematician in the entire British Isles. Having already been passed over for the chemistry chair, Herschel nonetheless traveled to Cambridge as soon as he heard the news of Milner's death to try for the late mathematician's position. He confided in Babbage that he had little hope for success but wanted to signal his desire and keep himself in the sights of the "big wigs of the university." Herschel was correct; this second attempt at a Cambridge chair failed. When the position became vacant again just months later, Herschel was too disgusted to make another attempt.[35]

Even though he was frustrated with Cambridge at an institutional level, Herschel continued lending his influence, energy, and support to reforming movements at the university. One of these was the Cambridge Philosophical Society, founded in 1819 by Herschel's friend Adam Sedgwick and former professor Edward Daniel Clarke. No doubt recalling meetings of the Analytical Society, Herschel contributed three papers to the Philosophical Society in its first years, all of which were published in the society's first volume of memoirs. These papers, which encapsulated his optical and mathematical work to date, were a chance to shape the work of the next generation of Cambridge natural philosophers, a group that included students like George Airy (1801–1892), who would later become an important astronomical collaborator, and the tutor William Whewell.[36]

In light of Herschel's continued Cambridge connections and his growing influence, upon his return to England his colleagues felt he was in a stronger position than ever to try once more for a permanent position at the university. They urged him, when the death of Samuel Vince (1749–1821) opened the Plumian Chair of Astronomy, to put himself forward for consideration. By this point though, apart from his frustrations with the university administration (he fumed he could "fairly calculate on *the permanence of ignorance*" of the decision-makers involved), his travels had altered his vocational desires. He no longer had "the keen relish for abstract mathematical studies" and was ready to leave the fight to reform mathematics to others. As far as natural philosophy was concerned, Herschel said he would "rather pass my days among those who are advancing eagerly . . . and running a race with ardor than in goading up hill the sluggish paces of any established institution under the Sun." Somewhere on those Alpine slopes, Herschel had decided he would contribute to natural philosophy on his own terms, as a traveler, observer, and experimenter, rather than as part of any conservative, traditional institution. Herschel was finished trying to reform Cambridge.[37]

Yet plenty of other institutions still needed reform. Herschel had been appointed a resident commissioner of the Board of Longitude, which put him in a position to address many of the Astronomical Society's complaints against it. Thomas Young, secretary of the board and superintendent of the *Nautical Almanac*, viewed the board's role in traditional terms: it existed to assist in matters of navigation. The board, Young believed, was primarily a nautical organization, not an astronomical one. Despite calls for change from people like Francis Baily and James South, Young felt that government money should not be used for purely astronomical pursuits. If improvements to the *Almanac* could not be tied directly to the board's original purpose—assisting with the determination of longitude—then they were unnecessary, regardless of whether they suited the desires of the new business astronomers. Herschel, as a close associate of both Baily and South (whom he was also observing with regularly), found himself in an awkward position. Baily urged on Herschel the importance of lunar occultations to be included in a revised and expanded version of the *Almanac*, while South pushed requests through Herschel that the board fund his equipment for double star observations. Herschel walked a tightrope of establishing himself as one of the board's junior members and navigating its relationship with the government while also trying to accommodate his friends' requests.[38]

Meanwhile, Herschel's astronomical work continued in earnest. His project with South measuring and comparing double star observations soon resulted in the publication of a catalog of hundreds of star pairs. This led Herschel to focus even more on instrumentation, continuing repairs and improvements to his father's twenty-foot reflector and for the first time using the massive forty-foot as well. His observations in London with South utilized South's refracting telescopes, and this along with Herschel's observations with his father in Slough meant he became well-versed in both types of instruments. In an observing community that almost exclusively used refractors, he was unique for observing with large reflectors, though his theoretical work focused on the optics of refractors due to the need to address chromatic aberration. This blend of theoretical and practical expertise meant that when the Royal Society formed a committee to develop better methods of glass manufacturing, he was an ideal candidate to serve on it alongside George Dollond and the natural philosopher Michael Faraday (1791–1867). With positions in the Royal Society, the Board of Longitude, and the Astronomical Society, and with international stature from his travels, Herschel had reached a position in London to begin reforming natural philosophy from within.[39]

DEATH OF AN ASTRONOMER

In July of 1822, Herschel's longtime friend James Grahame, who had recently recovered from an illness that kept him bedbound and barely able to write, invited Herschel on a tour of Holland and the Netherlands. Grahame, who had previously been uninterested in travel, said he had been inspired by Herschel's recent expedition with Babbage. Herschel was eager to spend time with his friend but was worried about being away for long due to his father's declining health. Grahame assured him their trip would be short, only ten days or a fortnight. "I go in search of nothing," Grahame told him, "but what can be got by a very superficial view." Instead of physical measurements and geology, this trip would be about architecture and the culture of the Low Countries, something that interested Grahame as an amateur historian. Grahame told Herschel he was willing to make the trip solo if Herschel could not join him, though he confessed he spoke not a word of Dutch and knew nothing of passports. He pledged to visit Herschel on his way back to Scotland.[40]

Perhaps it was concern for his friend traveling alone that convinced Herschel to return again to Europe. He tied up his affairs in London and invited Richard Jones to make the trip with them. Jones was once again unable, and Grahame and Herschel left London together for Dover on 8 August. In Calais they picked up the same carriage that had served Herschel and Babbage so well on their earlier travels and journeyed to the battlefield of Waterloo, arriving at sunset and exploring until dark. The road from Waterloo to Brussels was so poor it ruined the carriage that had survived Herschel and Babbage's trip through the Alps, and Herschel sold it in Brussels for, he complained, "a quarter of what I gave for it." In Brussels Herschel made camera lucida drawings of the city's imposing cathedral and visited the art galleries with Grahame. From there, they planned to travel to Antwerp.[41]

Back home, Mary Herschel wrote Herschel thanking him for the details of his trip he shared by letter and urged him to write more, telling him his father enjoyed his descriptions and was pleased he was enjoying his trip. William was, she assured him, as well as he had been when Herschel left. Unfortunately, as Herschel and Grahame continued their tour, his father's health took a sudden turn for the worse. On 20 August, Mary wrote a short note urging Herschel to return to Slough immediately. The letter did not reach him. Three days later, hoping he had returned, a relative wrote to Herschel's address in London to report that William was weaker but in no pain. Two days later, a Sunday, William Herschel died. Nearly a week passed, and there was still no sign of Herschel and Grahame.[42]

Herschel had continued his travels, unaware of the tragedy unfolding at home. He and Grahame did not return to London until 31 August. News came before their ship even reached the shore. Overwhelmed with grief and guilt, Herschel left Grahame in London and traveled immediately to Slough, where he found Babbage with his mother. Herschel had missed not only the opportunity to say goodbye to his father but his funeral as well, which had taken place while he was still abroad. Babbage had represented his absent friend among the mourners. Condolences and remembrances of the famous astronomer poured in from friends and colleagues across Britain and beyond, memorializing the observer who had opened the sidereal heavens and doubled the size of the planetary system. William was interred in a vault beneath his parish church of Saint Laurence in Slough on 7 September.

The world mourned an astronomer, but Herschel was alone in mourning a father. William had ensured his son the best mathematical education available in the kingdom and gave him an unequaled observational and instrumentational legacy. He had counseled his son on a career and vocation and had waited patiently while John at first resisted and then, slowly, began to put this astronomical heritage into practice. William taught John his methods, trained him in repairing the telescopes and forging their mirrors, and provided the resources to complete his life's work of surveying the entire heavens. But John recriminated himself that he not appreciated this and had not been sufficiently attentive to his father. He was sure that "a little forethought . . . on my part [would] have spared his kind heart many a pang."[43]

There was one obvious and inevitable way for Herschel to honor William's legacy. He immediately set about in earnest completing restorations to the twenty-foot telescope and preparing it for use. By 16 September, just over a week after burying his father, he had replaced the giant telescope's tube. Two days later, he took his place at the telescope again. In design, it was the same telescope William had used night after night sweeping the skies for objects never before seen. And yet in one sense it was an entirely new instrument. Herschel had, with his father, created and polished new mirrors, and most of the wooden structure had been completely rebuilt under his father's direction.

On his first night observing after William's death, as Herschel scanned a bright section of the Milky Way and saw the nebulous patch of light as clearly defined stars, he reinitiated his father's immense and incomplete survey of the heavens and mourned him in his own way: "The resolution is complete. The stars thick sown, like gold dust strewed on a black ground, but not running into each other. No trace of nebulous light between them. A truly magnificent sight."[44]

CHAPTER 5

GRAND TOUR

O thou Vesuvius whom I now survey
Not in the dreaming of a minstrel's eye
Not in the idle vision of a lay
But soaring cloud-clad through the native sky . . .

—JOHN HERSCHEL TO MARY PITT HERSCHEL, JUNE 1824

Vesuvius loomed in the distance. There had been a mild eruption the day before Herschel arrived in Naples, and through the windows of his hotel and across the bay he could see smoke escaping the volcano's summit. Vesuvius's peak was desolate, but its lower slopes were thick with vineyards that wreathed the volcano's shoulders. As he stared across the water at the smoldering volcano, the lines that came to mind echoed those of George Gordon, Lord Byron (1788–1824), another English traveler who had come to Italy in pursuit of a Romantic ideal. The sight of Vesuvius, the elegance of Naples, the sunlight on the blue waters of the Mediterranean, his entire journey south through Italy—all for Herschel represented the intertwining of his scientific pursuits with a search for artistic and poetic inspiration.

With the death of his father, Herschel became a rich man. William's will left his son twenty-five thousand pounds, a sum that gave Herschel immediate financial independence and made him the equivalent of a millionaire today. In 1824, three years after William's death, Herschel embarked on an extensive European grand tour, spending several months traveling through France, Italy, and Germany. Like his previous trip with Babbage, this voyage was undertaken in response to

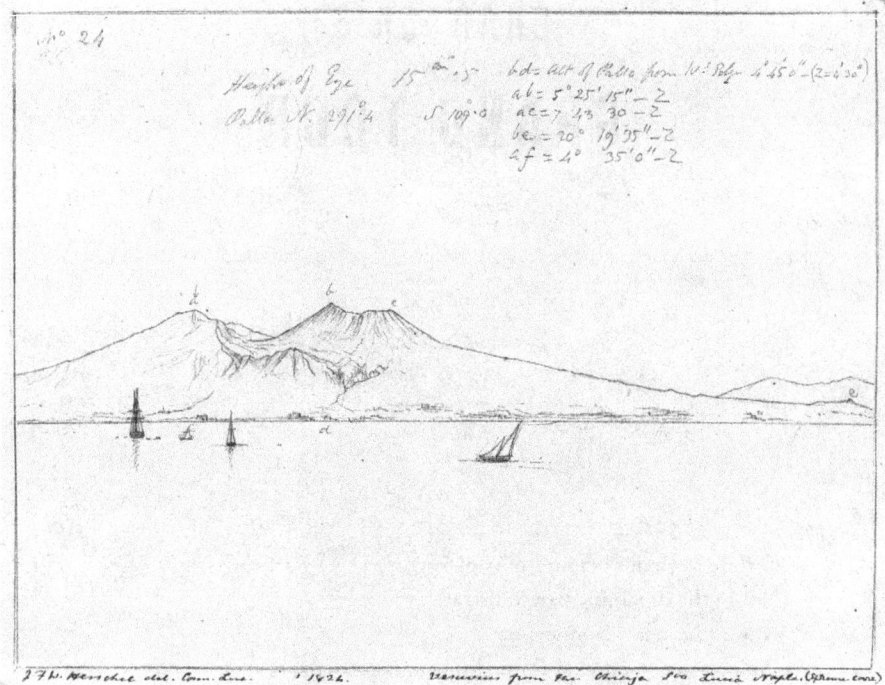

FIGURE 5.1: Sir John Frederick William Herschel, *Vesuvius from the Chiaja St. Lucia Naples*. Graphite drawing made with the aid of a camera lucida, unframed: 6 7/16 × 8 11/16 in. The J. Paul Getty Museum, Los Angeles, Gift of the Graham and Susan Nash Collection.

disappointment in England, this time of a professional nature, as his astronomical inheritance proved unequal to delivering Herschel from the intrigues of the London scientific community. Unlike his previous trips, though, this time Herschel traveled with only a servant and in search of historical and cultural as well as geological and astronomical sights. A Continental grand tour had long been part of the cultural heritage of young British aristocrats. By the time he embarked on a grand tour of his own, Herschel had established himself as a scientific authority. His tour added cultural credibility to his self-fashioning as an international polymath and savant.[1]

Herschel's travels extended and strengthened the networks of scientific practitioners he had been developing since the formation of the Astronomical Society. In particular, Herschel's travels in Italy and Germany brought him into contact with the leading opticians and producers of astronomical instruments outside London. What he saw confirmed the fears of his British colleagues that England was losing its leadership in the field. Instead of motivating him to pursue research on instrument construction, however, this knowledge would motivate Herschel's most expansive project of natural philosophy to date, a comprehensive treatment of optics and a survey of the scientific community's understanding of the nature of light.

The literal and figurative climax of his tour, as Herschel pushed south beyond Vesuvius and onward to Sicily, was his ascent of Mount Etna, Europe's highest volcano. Here, Herschel had his first glimpse of new skies. The constellations that appeared on the southern horizon were invisible from the skies over Slough and included some of the richest regions of the Milky Way. These new zones of clustered stars and nebulosity planted the seed in Herschel's mind to extend his father's telescopic surveys to encompass the entire heavens. From the peak of Etna, watching the sun rise over the Mediterranean, Herschel began to consider taking his conception of science to the farthest fringe of the British empire and summiting the grandest astronomical peak of all.

Arriving in Naples, as Herschel gazed across the bay toward the slopes of Vesuvius, he could not have realized the tragic irony of the Byronic lines he echoed. Even as Herschel traveled southward through Italy, in search of his own Romantic ideal, Byron had died of a fever contracted at the siege of Missolonghi leading Greek forces in their war of independence. The poet had already become an exemplar of British engagement with nature and the classical world that characterized Romanticism. But whereas Byron emphasized a suspicion of the physical sciences and industry, Herschel offered a synthesis of the natural and cultural world as an artist seeking sublime experiences and a natural philosopher studying the physical landscape.

Herschel's rhapsodic responses to Vesuvius, observations on Florentine art, and meticulous drawings of Roman ruins were made alongside sketches of geological strata, temperature records from the four carefully calibrated thermometers he carried everywhere he went, continuous barometric observations, and notes on the astronomical observatories of each city visited. For Herschel, in contrast to Byron—and more importantly for the scientific sensibility Herschel was fashioning for himself—this was exactly the point: it was not poetry *versus* natural philosophy; it was not a tension between physical measurement and artistic sensibility. Rather, Herschel's grand tour showed that only the synthesis of the emotional and scientific ideals represented a true, complete experience of the world.

AN ASTRONOMICAL LEGACY

The death of William Herschel left his son with a vast inheritance, both financially and astronomically. William had, with his sister Caroline, populated the cosmos with new sidereal objects: double stars in gravitational orbits, vast nebular clouds congealing to form stars and planetary systems, and immense stellar clusters of uncertain size and extent. His work offered a vision of a dynamic universe extending beyond the solar system. With the death of his father, John Herschel began to feel that his role would be to establish this new form of astronomy by making William's observations useful and accessible to the wider astronomical community. William's catalogs of nebulae and double stars were difficult to use, with their contents organized not by position of objects in the sky but rather by order of discovery. In addition, the location and description of William's objects were not precise enough to allow other observers to find them, observe them, and establish an empirical baseline on which to compare later observations. And John Herschel was certain that there would be later observations. His colleagues in the Astronomical Society, many of them with the financial and industrial resources to build large telescopes, would soon be revisiting his father's double stars and nebulae. Herschel believed he could pave the way by revisiting William's observations and revising his earlier catalogs, a project he had already begun for double stars with James South.

But double stars were just one part of the Herschellian legacy, and despite his father's death, Herschel still observed only occasionally with the large reflectors at Slough. In 1823 it seemed to Herschel that the best way to carry on his father's astronomical legacy might be not under the open skies of Slough but rather from the observatory dome at Greenwich, as it began to appear he would be made the chief astronomer in the kingdom. The astronomer royal, a position created in 1675 with the founding of the Greenwich Observatory, was Britain's most prestigious astronomical role. It was also, beyond the two chairs in astronomy at Cambridge and Oxford, the only "professional" position for an astronomer in England. And whereas university chairs carried with them the institutional baggage that had frustrated Herschel at Cambridge, the position of chief observer at Greenwich would allow him to set the program for astronomy throughout Britain. As astronomer royal, Herschel would be able to establish his father's sidereal astronomy alongside the more traditional forms practiced at Greenwich for more than a century.

The current astronomer royal was John Pond (1767–1836), who by the start of 1823 was in poor health and not expected to live much longer. Pond had been nominated to the post after the death of the previous astronomer royal, Nevil Maskelyne, who had been highly respected for the quality of his positional star observations. Pond was the nominee of Joseph Banks and a way for Banks to retain influence over Greenwich. Though Pond was respected as an astronomer, his organizational and administrative skills were not up to managing the production of the *Nautical Almanac* and overseeing the calculators who reduced Greenwich observations. Because of this financial and organizational disarray, oversight of the *Almanac* had in 1818 passed from the astronomer royal to the Board of Longitude.

Herschel was an obvious candidate to replace Pond. He had the credibility, experience, and motivation to restore Greenwich's efficiency and steer its work along the lines most useful for the new generation of astronomical businessmen in the Astronomical Society. Unfortunately, the person with the most influence in choosing Pond's successor was the new president of the Royal Society, Humphry Davy, and into this delicate situation stepped South. South had Davy's ear, but Herschel's astronomical collaborator seemed incapable of delicacy, tact, or correctly understanding Herschel's intentions. Davy, who no doubt recalled Herschel's strident opposition to his presidential candidacy, nonetheless recognized that Herschel was the best candidate for the role. In the meantime, though, there was the question of appointing a chief assistant for the ailing Pond. South felt so strongly about Herschel's chances of becoming astronomer royal that he wanted Herschel's opinion about potential assistants, with the assumption that whoever was appointed would soon become Herschel's assistant.

Herschel agreed it was probable he would be offered the position of astronomer royal and admitted that if nominated he would be *"anything but disinclined."* He took South's questions about a suitable assistant seriously, especially as the appointment would represent a significant change in how Greenwich had been managed in the past. Previously, assistants had been observers with little additional training. But Davy was considering appointing educated gentlemen who would have greater autonomy in decisions and would help administer the observatory rather than simply carry out orders. This, Herschel emphasized, meant the character of potential assistants became all the more important, but he felt that Pond was not discerning enough to nominate suitable help. Being a competent observer did not make one a virtuous person. Herschel worried about an assistant of "good talents" with "astronomical and mathematical knowledge more than competent" who was nonetheless "unprincipled and bankrupt in character, devoid of common honesty . . . un-trustworthy to the very last degree, and in his habits always calculating to make a case for himself at the expense of others, impertinent in thrusting forward his own suggestions, envious of any merit but his own and with strong principle of enmity in his disposition." Herschel seemed to have more than a hypothetical case in mind. He was worried about someone in particular. If this warning was not clear enough, Herschel ended his letter to South on this topic with a postscript urging that no appointments be made at the time, obviously hoping to make the decisions himself when he replaced Pond.[2]

South relayed Herschel's thoughts to Davy, but to Herschel's frustration South appeared to have miscommunicated his intentions. From Herschel's emphatic response after South recounted his conversation, it's clear that Herschel was worried about a specific individual becoming head assistant: one Stephen Lee (d. 1835), a fellow of the Royal Society and an assistant secretary since 1810. Herschel, who was generally easygoing enough to be able to get along with short-tempered characters like Babbage and South, had no intention of working with Lee in any capacity and expressed his *"positive resolution to have nothing to do with any transaction* which may render it necessary . . . to have frequent and confidential communication with a person like Stephen Lee." Herschel insisted he did indeed "covet with earnestness" the position of astronomer royal but not at the expense of having to work with Lee. (Herschel's evaluation of Lee was eventually validated. After Lee launched a "virulent criticism" of Greenwich's observations, a subsequent investigation led to his resignation from the Royal Society.)[3]

In the end, the discussion was moot. Pond would live another twelve years and serve as astronomer royal until shortly before his death in 1835, and there is no indication Lee was ever appointed an assistant at Greenwich. But the exchange colored Herschel's view of pursuing astronomy in a professional capacity. Science, including astronomy, was a virtuous endeavor, dependent on the qualities of its practitioners, and institutional independence was the only foolproof means of preserving this virtue. From this point on, Herschel began to look for ways to maintain his own independence and pursue his projects without official entanglements. "I mean to make Science my amusement," he informed South after he had achieved some distance from the Lee debacle and had time to reflect more on what the administration of Greenwich would have entailed, "not my business."[4]

Herschel still retained a measure of institutional involvement in his roles in the Royal and Astronomical Societies and as a member of the Board of Longitude, but he was beginning to see the institutional organization of natural philosophy and its practice as two separate things. More and more, Herschel perceived the bodies intended to support and promote the sciences as distracting and stymying their actual progress. The work of his close colleagues seemed to illustrate this. On the one hand he saw South and Francis Baily in vociferous public rows over how astronomy should be conducted and supported; on the other, he saw Babbage wrangling with the government for funding of the construction of his new calculating machine. Herschel, with the new financial resources afforded by his inheritance, resolved on a path of personal independence. He would ultimately follow this path to the Cape of Good Hope in South Africa. But in the immediate wake of his frustrations over the position of astronomer royal, he decided to again depart for the Continent, this time on an extensive European grand tour.

GRAND TOUR

By April 1824, when Herschel once again journeyed to Europe, he was thirty-two years old and a seasoned traveler. But this time, instead of traveling with friends as a young mathematician or seeking solace from heartbreak, he traveled as a moneyed gentleman with a servant named James Child and two hundred pounds of luggage. As always, though, he went with the recognition afforded by his family name. At the customs house in Calais, for instance, when officials saw his passport they noted that it bore not the name of a person but rather "le nom d'un planete" and let all his instruments pass without inspection. Over and over during this trip, Herschel would call at observatories or the homes of leading men of science without introduction, only to be immediately welcomed and treated as a distinguished guest.[5]

Herschel was traveling in the footsteps of many moneyed English travelers. A tour of the Continent and the historic sights of Italy had long been seen as a way of obtaining a veneer of cultural polish. But a growing number of savants were making their grand tours as geologizing expeditions. George Bellas Greenough (1778–1855), who would go on to become the first president of the Geological Society of London, had traveled from Chamonix in the Alps to Vesuvius in southern Italy during the Peace of Amiens, when Herschel was a boy visiting Paris with his father, and in 1816 Greenough traveled extensively in Europe with Oxford geologists Buckland and Conybeare and then again in 1820 with Buckland. Herschel's own grand tour would combine geologizing, connecting with European astronomers, and indulging his own poetic sensibilities. He had worked out specific guidelines for travel, recording them in one of the travel notebooks that he carried with him and reflecting his determination to make the most of his time and record everything he saw:

1. Keep a book like this & a duplicate and *use it*.
2. Economy of Time is *Economy of Money*.
3. Keep your journal and *your temper*.
4. Put your pride in your pocket, and take out *your money*.
5. Never be frightened by *bugbears*.

5 ½. One look on the spot will prevent a thousand longings at a distance.

5 ¾. One *line* on the spot is worth a thousand recollections.

6. Be communicative.

6 ½. Never cut a figure.

7. Avoid travelling at night and in *bad weather*.

8. Despise petty *impositions*.

9. Beware of *Malaria*.

10. Shock no *prejudices*.

11. Make points.

12. Be never an hour without an object.

13. In bad weather bring up your journal.

14. Do not make yourself a slave to your instruments.

15. Enjoy whatever is enjoyable as what you may never enjoy again and overlook whatever is disagreeable as what may never happen again.

Most of these he had clearly learned from experience.[6]

In Paris Herschel reconnected with colleagues, though he had no intention of lingering in the city long. He attended a session of the Académie with François Arago and was introduced to chemist Eilhard Mitscherlich (1794–1863). He also visited the Paris Observatory, where he examined the instruments and conducted observations, and he spent a morning working in Jean-Baptiste Biot's laboratory, comparing the results of their various optical experiments. Alexander von Humboldt called at his hotel, where Herschel quizzed the explorer again regarding volcanoes. Herschel had pledged to take no dinner invitations, fearing if he did he would be in Paris for weeks satisfying social obligations, but he was convinced to attend a concert at the home of the chemist Louis Jacques Thénard, only to have André-Marie Ampère (1775–1835), discoverer of electromagnetism, talk so loudly and excitedly to Herschel about his work that no one could hear the music, much to the amusement of the other guests.[7]

But Paris this time was only an obligatory stop, and Herschel was anxious to press onward. From Paris he retraced his earlier route with Babbage, passing through the Alps in the midst of a heavy snowstorm. The weather soon cleared, and by the time he arrived in Turin, Herschel's hands and face were severely sunburned from the mountain passes. His burns reminded him of how Babbage had suffered similarly after their ascent of Mount Rosa. Clearly, sunlight was more intense at higher altitudes, and this gave Herschel an idea for an additional instrument to add to his arsenal: a device for measuring the intensity of solar radiation, which he would eventually call an actinometer. Along with his thermometers and barometers, the actinometer became central to Herschel's attempts to measure properties of the atmosphere at high elevations.

The actinometer, which he had crafted to his specifications in Turin, was an altered thermometer comprised of a cylinder of dark liquid affixed to a tube with precise gradations. Whereas a thermometer uses a liquid's expansion to measure the temperature of the air, with the actinometer Herschel observed the difference in the expansion of its fluid in the shade and then in sunlight, thus measuring not the temperature of air but rather the intensity of solar energy absorbed by the liquid. By noting how much greater the absorption of solar energy was at a mountain's peak versus at its base, Herschel sought to determine how solar energy varied with elevation and thus how much of the sun's energy was absorbed by Earth's atmosphere. Taking an actinometer to the top of a mountain, Herschel realized, was the best way to gain estimates of how much solar radiation was incident on Earth, a key to understanding weather systems and long-term changes to Earth's climate.[8]

In Turin Herschel met Giovanni Plana, director of the Turin Observatory, and, as he had at Paris, carefully examined the quality of the instruments in Plana's observatory. This evaluation of observatory instruments throughout Europe became a theme of Herschel's tour, and he stopped in as many observatories along his route as possible, observing with the instruments when there was opportunity and clear nights. This strengthened the network of astronomers he had begun to establish as foreign secretary to the Astronomical Society, but it was also part of his efforts to determine whether the concerns of his colleagues in England were well-founded. London instrument makers were worried that British preeminence in manufacturing was waning, and Herschel wanted to see whether this was true.

During William Herschel's lifetime, every notable European observatory had been equipped with instruments by British manufacturers. Now, instrument makers like Edward Troughton and George Dollond, friends of Herschel's and members of the Astronomical Society, feared instruments from European makers were surpassing their own. Rumor had it that the glass manufactured by the Bavarian Joseph von Fraunhofer (1787–1826) was far superior to anything the British could produce. Herschel also learned of an astronomer in Italy, Giovanni Battista Amici (1786–1863), who was making reflecting telescopes and micrometers of his own design. To examine these instruments, the free access provided by the Herschel name became incredibly useful. When Herschel arrived in Turin, for instance, though he had never met nor corresponded with Plana and lacked even a letter of introduction, he was received warmly and spent his five days in the city observing with Plana each evening. As he would learn, Troughton and Dollond indeed had reason to be concerned.[9]

From Turin Herschel traveled south to Genoa on the Mediterranean coast. By this point in his tour, Herschel's notes, which had focused largely on astronomy and the geology of the landscape through which he passed, begin to include extensive observations on architecture and art. Herschel visited the churches and galleries of Genoa and was struck by the city's sad state of decay. Many of the libraries of the city's nobles were being auctioned off. Herschel visited the Durazzo Palace (now the Palazzo Reale) to see about making purchases from its library but instead found the building closed and the Sardinian prince sheltering within for fear of political turmoil. Herschel wrote to Babbage that the condition of the Kingdom of Sardinia, which at this time ruled Genoa, made him thank God he was an Englishman.[10]

In Genoa Herschel met Baron Franz Xaver von Zach (1754–1832), a Hungarian astronomer best known for establishing the first regular astronomy periodical, the *Monatliche Correspondenz*, which was published from 1800 to 1813. At the time Herschel met him, Zach was editing a new, similar periodical, the *Correspondence Astronomique*. Herschel found Zach "a most amusing old Gentleman full of anecdotes and the greatest Gossip" though somewhat of a "scandal-monger." He was dismayed to learn that Zach was feuding with the French astronomer Jean-Louis Pons (1761–1831), a situation that put Herschel in an awkward position, as he had been charged with presenting Zach a medal from the Astronomical Society for Pons for the latter's cometary discoveries. As with the arguments between Biot and Arago on his previous trip to Paris, Herschel was frustrated to find contention and dispute where he hoped to see cooperation.[11]

From Genoa Herschel traveled to Modena, where he met the astronomer Giovanni Amici. In Amici Herschel found a kindred spirit. Alone among astronomers during this period, Amici was following William Herschel's footsteps and creating large reflecting telescopes that John Herschel believed were the best in Europe. The Italian astronomer was also the first observer Herschel had encountered outside of England conducting double star observations. Amici, influenced by the elder Herschel's work, had begun making systematic observations of double stars even before Herschel had begun his own work with South. Herschel spent an evening observing with Amici, using the Italian astronomer's micrometer to measure double stars. When Herschel left Modena, he offered to buy one of Amici's micrometers, only to have the astronomer give it to him as a gift. Amici's work confirmed Herschel's suspicions regarding the increasing quality of European instruments and made it clear to him that English instrument makers would have to work hard to keep from being left behind.[12]

Observatories and their instruments were only one aspect of the huge amount of observations Herschel recorded on his tour. He constantly noted and interpreted what he saw as he traveled. Not a mile of road passed without observation. Herschel commented on the appearance and disposition of peasant beggars, the architecture of each town he passed through, and the geological strata and minerals found along the roadsides. He recorded atmospheric pressure, temperature, and weather, what kind of clouds were in the sky, how the tops of mountains appeared, whether they were snow-clad or cloud-capped, and what this indicated about temperatures at high altitudes, keeping up this consistent level of observation as he continued southward.

From Modena Herschel traveled to Bologna and from there crossed the Apennines (which his servant James referred to as the "Happy Nine Hills," a name he attributed to the "Nine Musics"). Along a precipitous mountain road between Bologna and Florence, one of his coach's horses faltered on the road's steep edge. The driver had time for only a shouted "My God!" before the entire wagon toppled over. Herschel, sitting on the opposite side of the coach, was launched upward as it overturned. The day was fair, so the top of the carriage was down. Otherwise, Herschel reflected afterward, he would have been carried down the slope with it. As it was, he saw the horses break away and James, who had been riding in front, disappear between them as he was hurled down the slope. When Herschel landed, he was sure the overturned carriage was tumbling down the mountainside to crush him, and he scrambled out of the way. He was not quick enough, and time seemed to slow as he watched the coach pass over his head as it fell, coming to rest several hundred feet farther down the slope.[13]

The accident left Herschel badly shaken. If it had happened a hundred yards earlier or a hundred yards farther on, where there were steep precipices along the road, Herschel was sure he would have been killed. As it was, the postillions gathered the horses and found the carriage itself surprisingly undamaged. Herschel got hesitantly to his feet and was amazed to see James emerge from the tangled harnesses unharmed as well. With the help of some passersby, they pushed the coach back onto the road and were under way again within the hour. Herschel's deliverance, he recounted afterward from the safety of the next inn, he could "hardly regard as short of a miraculous interposition."[14]

They did not emerge completely unscathed: the accident destroyed some of Herschel's instruments, including one of his precious barometers, but it did not stop him from a planned detour on the way to Flor-ence to view local volcanic vents, where he found hydrogen, flames, and evidence of past lava that had seeped up from an area about four or five yards in diameter on a limestone hill. Herschel sampled rocks, sniffed gases, and blocked vents with his hands, amusing his guides and terrify-ing James. It was his first glimpse of the volcanic activity he was seeking in southern Italy and that he would witness on a much larger scale at Vesuvius and Etna.[15]

Florence, with its renaissance art and architecture, was the jewel of the traditional grand tour, and Herschel was not disappointed. He viewed the city's sculptures, including the famous Venus de' Medici, and spent two mornings in Florence's art galleries before traveling to the garden of Pratolino to view the Colossus of the Apennines, a huge statue by Giambologna (1529–1608). He also visited the monastery of Vollambrosa, where he paid his respects not as a philosopher but as a poet. Vollambrosa had become important to English travelers because of its mention in Milton's *Paradise Lost*, where Milton refers to fallen an-gels lying "thick as autumn leaves that strew the brooks in Vallombrosa." Herschel saw no leaves of autumn, traveling in the heat of summer, but he paid homage to Milton with a stay in the ancient monastery. Poetry was a part of Herschel's identity as well: three short works of his had been collected in an anthology the year before his travels. Vallombrosa itself Herschel found "hardly worth the trouble of a drawing or descrip-tion," though the monks were as kind as could be expected from people whose livelihood depended on their hospitality.[16]

Herschel's poetic sense was more satisfied by ruins adorning gardens or viewed by moonlight. When Herschel reached Rome, however, he found the splendor of the past alongside contemporary decay, making the Eternal City seem dismal and decrepit. The forum, Herschel was disappointed to discover, was now a cow-market filled with rubbish, the palace of Augustus a series of barns and dunghills. Only the splendor of Saint Peter's and the majesty of the Coliseum, which Herschel spent two days sketching with the camera lucida, lived up to his expectations. South of Rome, the country of the Papal States seemed a wasteland, with grassy plains broken by ruins and huge aqueducts that showed the extent and engineering prowess of the ancient Romans. At the border with the Kingdom of the Two Sicilies, which controlled the land around Naples as well as the island of Sicily, the landscape changed drastically: grasses gave way to aloe and cactus, fields to orchards of lemons and oranges. In the country surrounding Naples, poplars stood among fields of corn, and grape vines were trained up the trees so that the countryside seemed a combination of field, forest, and vineyard.

FIGURE 5.2: Sir John Frederick William Herschel, *Interior of the Roman Colosseum, May 18, 1824*. Graphite drawing made with the aid of a camera lucida, unframed: 7 15/16 × 12 3/16 in. The J. Paul Getty Museum, Los Angeles, Gift of the Graham and Susan Nash Collection.

Herschel, like many British visitors of the previous century, was enchanted by Naples and its view of the Mediterranean. In contrast to Rome, everything in Naples was elegant, from the lodgings he and James found to the way peasants in the street addressed him as "your excellency." Herschel immediately connected with the Royal Academy of Naples, the city's scientific society, and introduced himself to Teodoro Monticelli (1759–1845), the academy's secretary and leading authority on Vesuvius. Monticelli was the center of a network of Italian natural philosophers and had guided other visitors on previous expeditions to Vesuvius. Humphry Davy, for instance, had visited the volcano in 1815 and 1820 to test his theories on the origins of volcanic heat. Likewise, when the geologist Charles Lyell (1797–1875) followed Herschel's footsteps in 1828, Monticelli provided details on the mountain though it was impossible to visit then due to eruptions. Vesuvius, as the most active and accessible volcano in Europe, was the ideal place to observe volcanic activity and test theories on the formation of rocks and minerals.[17]

Vesuvius had famously erupted in AD 79, burying the ancient city of Pompeii and creating pristinely preserved ruins that were excavated

FIGURE 5.3: Sir John Frederick William Herschel, *Interior of the Crater Vesuvius, June 9, 1824*. Camera lucida drawing on paper. Gift of Susan and Graham Nash, courtesy of the Museum of Photographic Arts, San Diego, 2002.040.035.

by the time of Herschel's visit. In the 1600s, Vesuvius began a period of activity that continued with relatively severe eruptions throughout the eighteenth century. There was another major eruption in 1822, visible for miles out at sea and dramatically captured in many contemporary images. Such activity and accounts made Vesuvius loom large on both scientific and Romantic cultural landscapes. Natural philosophers in particular debated whether the volcanic heat so dramatically illustrated by Vesuvius's activity was caused by violent chemical reactions or by Earth's internal temperature. Volcanoes were also important for understanding the origin of minerals like basalt and whether they had precipitated from a global ocean or were the results of igneous processes. George Julius Poulett Scrope (1797–1876), like Herschel a graduate of St John's in Cambridge, had been present at the 1822 eruption and went on to outline a theory of volcanoes based on his observations. Herschel, like Scrope, suspected volcanism was caused by internal heat, with volcanic eruptions triggered by rocks expanding due to heat, though at this time it was unclear whether the temperature inside Earth continued to increase with depth. Vesuvius was central to answering these questions, and contemporary understanding of volcanism was based largely on observations of and experiments on this single volcano.[18]

After a week in Naples, Herschel was ready to set off with James and the Neapolitan chemist Nicola Covelli (1790–1829) for the volcano. They followed a well-worn path trod not only by natural philosophers but also by writers like Percy Bysshe Shelly (1792–1822), who had climbed the volcano in 1818. The same ascent would be made by Charles Dickens (1812–1870) and the art critic and meteorological enthusiast John Ruskin (1819–1900) in years to come. It was a climb much less arduous than the Alpine ascents Herschel had made with Babbage, requiring only a walking stick and a pair of boots. Herschel and his companions left for the volcano in the evening with the intent of climbing by torchlight and watching the sunrise from its summit. Rain delayed them at the foot of the volcano until 4 a.m., when they began the climb with donkeys and mules, passing through vineyards and breaking often for barometrical measurements and for Herschel to make notes on the structure of solidified lava flows. At a shelter at the foot of the volcano's upper cone, they left the animals and proceeded on foot, climbing slopes of rock, debris, and ash where their feet sank to the ankles with each step, to reach the rim of the crater.[19]

Looking down the hundreds of feet into the crater itself, Herschel could see smoke issuing from holes and fissures but no means of descending into the caldera. Unfortunately (or perhaps fortunately, considering how close Davy had come to asphyxiation while observing one on his previous visit) there were no active lava flows for Herschel to witness. As Covelli set up chemical experiments at a volcanic vent, Herschel traversed the crater's edge to its highest point. Suddenly, with a sound he compared to a thousand coppersmiths hammering combined with hundreds of bells all out of tune, a portion of the crater wall opposite slid into caldera. Herschel stayed at the summit long enough to make a careful drawing of the crater's interior before finally descending, much to James's relief.[20]

Besides Vesuvius, Herschel's time in Naples also included sailing expeditions on the bay and a trip to Pompeii, where Herschel saw the perfectly preserved city and imagined what it would be like to close up an English town for seventeen centuries before unearthing it. Perhaps the most significant visit, though, in terms of Herschel's evolving geological ideas, was to the temple of Serapis in the nearby town of Pozzuoli (since reidentified not as an ancient temple but rather a columned market). The violent eruptions of Vesuvius and the ruins of Pompeii illustrated the power of cataclysmic geological events to reshape the landscape. The question that geologists were wrestling with was whether such cataclysms were indeed the primary forces at play in Earth's history. Were changes on Earth's surface driven by violent upheavals, as geologists like

Buckland argued and the fossil researches of Cuvier suggested, or by the gradual accumulation of slow changes, as the followers of Hutton maintained? The ruins at Pozzuoli provided an important clue.

Serapis had become a centerpiece in this debate since Andrea de Jorio (1769–1851), an antiquarian living in Naples, had published a guidebook on the region that featured the temple. The ruins of Serapis contained a line of three upright marble columns marked to a height of ten or fifteen feet by perforations from a type of marine shellfish. This showed that at one time the columns had been submerged to that depth in the sea. Since the pillars were originally constructed on dry land, in the centuries since they had been erected the landscape had apparently seen a significant decrease and then subsequent increase in sea level. And because the pillars were still upright on their bases (Herschel checked them with a level and found them quite straight), this change could not have been caused by an earthquake or other rapid cataclysm. The ruins at Pozzuoli, Herschel could see, provided powerful evidence supporting the view of slow, steady, and uniform processes shaping Earth's surface. This example was so strong that when Lyell later published his seminal work promoting a uniformitarian view of geology, which became a paradigm of modern geological science, the frontispiece to his book was a copy of the engraving of the columns at Serapis. It was Herschel who urged Lyell to visit them.[21]

Herschel's original plan had Naples as the southern culmination of his travels, as the city and its famous volcano were the usual end of the line for the traditional European grand tour. These plans changed after a dinner where Herschel met the admiral of the British forces in the Mediterranean and heard detailed accounts of the sights of Sicily, which sounded appealing enough he decided to continue to the island. Sicily offered more ruins and geological sites, including Mount Etna, a much larger volcano less accessible and far less studied than Vesuvius. Getting to the island had recently become much easier as well, with the commencement of a new steam packet line between Palermo, the island's chief city, and Naples. Herschel witnessed the first arrival of the steamship into the Bay of Naples, and he would be among the passengers on its maiden voyage to Sicily.[22]

Finally and perhaps most importantly, Sicily offered a chance to visit Europe's southernmost astronomical observatory in the city of Palermo. Here, observers created stellar catalogs that included stars only visible from this far south. The farther Herschel ventured, the more new stars appeared above the horizon. The appeal of the classical world along with his abiding interest in geology had drawn Herschel to the edge of Europe and now compelled him to continue his voyage, where the most important vistas that opened to his view would be those in the skies above. In the nights of the southern Mediterranean, Herschel would glimpse for the first time skies his father had never seen.

SOUTHERN HORIZONS

The voyage between Naples and Palermo via steam packet was almost exactly twenty-four hours, and Herschel stayed on deck the entire night, watching southern constellations swing higher into the sky than he had ever seen them from England. For the first time, Herschel saw the brightest part of the Milky Way, which was hidden below the horizon in Slough. This region, thick with uncharted nebulosity and star clusters, would continue to draw his eye during his nights in Sicily. The ship's arrival the next day caused a sensation for the inhabitants of Palermo, filling the shore with cheering crowds and the harbor with local boats to escort the steamship into port for the first time. The effect of the ship's arrival was heightened when a two-masted brigantine, which had left Naples almost two days before, sailed into the harbor an hour and a half after the steam packet's arrival.

Steaming into Palermo, Herschel found the city's setting, ringed like an amphitheater by the surrounding mountains, even more impressive than that of Naples. For the first decade of the nineteenth century, Sicily had been ruled as a British protectorate, a bulwark against Napoleon's expansion into the Mediterranean. After the French defeat, Sicily, along with Naples, became independent as the Kingdom of the Two Sicilies but remained closely allied with Great Britain. Though the arrival of this first steam packet meant a closer connection between Sicily and the mainland and would soon allow more travelers, at the time of Herschel's visit the island remained a backwater with poor roads and difficult travel that Herschel's servant James was convinced was the end of the Earth. Even Herschel had to admit that it felt much like it.[23]

In Palermo Herschel called on Giuseppe Piazzi (1746–1826) at the palace observatory. Piazzi, like Herschel's father, was the discoverer of a planet: Ceres, discovered in 1801 and now known as a dwarf planet but previously considered the first of the minor planets, or asteroids, found between the orbits of Earth and Mars. The Sicilian astronomer had made his own scientific pilgrimage to England years before, where he visited Slough and met with William Herschel shortly before John's birth. At first, the seventy-eight-year-old astronomer took John for his half brother Paul, his mother's son from her previous marriage who had died when John was still a child. Once he realized who John was, he asked after his mother and aunt's health, the telescopes at Slough, and how John was continuing his father's work. Piazzi was no longer actively observing and had passed responsibilities to his assistant Niccolò Cacciatore (1780–1841), with whom Herschel made barometric measurements on the slopes of Monte Cuccio overlooking the city.

After spending time in Piazzi's famous observatory, Herschel was eager to begin his expedition into the Sicilian countryside. The roads were so bad throughout the island that wheeled vehicles were not used outside Palermo, so Herschel, James, and their local guides traveled entirely by mule, making thirty to fifty miles a day between villages and ruins strung along a largely unpeopled and rugged landscape. They traveled in a climate Herschel found striking for its absence of green plants, with the heat of summer broiling them by day and the fear of malaria haunting them at night. Though Herschel managed to convince James they had not actually reached the edge of the world, it did indeed seem to Herschel "a new world," a landscape of cloudless skies, spectacular sunsets, sunburned hills, thick orange groves, hedgerows of tall spiky aloes, and thickets of enormous cacti. Most of the island was arid, with only an occasional sulfurous and brackish spring.[24]

For all the difficulty of traveling, Sicily's features made the endeavor worthwhile. The island was dotted with ruins that dwarfed everything Herschel had seen so far, all in a setting quite removed from the urban squalor of Rome. The ruins of Sicily, which included the temple of Segesta and the Greek colonies of Selinunte and Agrigento, were immense and isolated reminders of the island's previous splendor. Herschel was struck by the picturesque setting, where the ruins were bathed in silence and felt utterly abandoned by the outside world, as well as the scale of engineering they represented. At Selinunte, for example, he was stunned to measure a column still standing that was thirty-one feet in circumference. A capital of the same temple was composed of a single stone square of thirteen feet on each side. As impressive as this was, there were not one but three temples of similar size. "It is indeed the beauty of desolation," he wrote, "for finer architectural remains can hardly be imagined." Mineralogically and geologically Sicily was equally fascinating, though Herschel found that the heat and difficulty of transport limited the number of samples he could collect and carry. Nonetheless, at the sulfur mines at Cattolica Eraclea, despite the fact that hammering in ninety-five degrees made him quite selective, he managed to gather too many rock specimens for the mules to carry.[25]

By the beginning of July, after a ten-day journey of two hundred miles across the rocky Sicilian terrain, Herschel and his companions reached Catania, at the foot of Mount Etna. Like Vesuvius, Etna would be both a Romantic and scientific conquest for Herschel as well as another opportunity to test his theories of volcanic heat against observa-

FIGURE 5.4: Sir John Frederick William Herschel, *Etna—the summit called the "Bicorne" from Casa Inglese*. Graphite drawing made with the aid of a camera lucida, unframed: 7 3/4 × 11 5/8 in. The J. Paul Getty Museum, Los Angeles, Gift of the Graham and Susan Nash Collection.

tions. Unlike Vesuvius, though, Etna was not an easy day's climb from the comfort of Neapolitan elegance. At two and a half times the height of Vesuvius, Etna was the largest active volcano in Europe, with documented eruptions stretching back into antiquity, long before Vesuvius became active. It had released a flow of lava that had rolled past the walls of Catania in 1669, reaching as far as the ocean, and had been in a nearly constant state of minor eruptions since, yet it also boasted lava flows that appeared far older than human records. Etna, Herschel had decided, would mark the ne plus ultra of his grand tour.[26]

After preparations in Catania, Herschel set off with James and their guides up the volcano's slopes, passing through forests covering its lower reaches and circumventing the scars of old lava flows to reach a hut built to shelter travelers near the summit. Along the way, Herschel compared the lavas of Etna with those he had seen at Vesuvius. What he saw here, he believed, indicated the greater depth of Etna's volcanic activity. Minerals and rocks thrown up by Vesuvius seemed more diverse, whereas Etna primarily produced lava of a more uniform consistency. This meant, Herschel suspected, that the subterranean fires powering Etna's eruptions were located below the intervening rocks. Herschel also obtained an important series of barometric readings on the ascent, coordinating his observations with Sicilian colleagues back in Catania and at a village ten miles up the volcano's slope and helping determine the precise height of the volcano, as he had for his climbs up the Breithorn and Staubbach Falls.[27]

Vesuvius had been ascended and descended in a single day, but the height of Etna meant that Herschel and the others passed the night near the volcano's summit. From the "highest house in Europe," Herschel composed letters to Babbage and his aunt Caroline, both figures with symbolic importance for the climax of his European grand tour. In his letter to Babbage, the analytical mathematician who had been interested only in the heights of logical formalism was nearly gone, replaced by a Romantic explorer eager to interpret the entire world—sky, heavens, and subterranean fire—through careful and extensive observation. Herschel had come a long way from the goals he had once articulated with Babbage for reforming British mathematics, moving toward a view of natural philosophy that used those mathematics to embrace the physical world. It is unlikely that Herschel the Cambridge mathematician would have recognized Herschel the philosopher who perched at Etna's summit, writing to Babbage as he awaited sunrise.

Caroline Herschel would have been pleased with her nephew's perspective from Etna, for on the peak of the volcano the southern skies continued to draw Herschel's eye and imagination. He had carried a small refracting telescope on his travels and used it to examine the unknown portions of the Milky Way now visible, noting how it was broken into clusters and nebulosity. Everywhere Herschel went, he had been greeted on the strength of his father's name, and now here at the southern limits of Europe, Herschel glimpsed stellar vistas that were beyond anything his father had surveyed. There were, Herschel was perceiving, entire new regions of the sky to explore. He had ascended the Alps and the volcanoes of Italy, but now he caught sight of the highest summit remaining to scale: returning to William's surveys of nebulae and expanding them to the entire heavens. This would require traveling even farther abroad. It would also, Herschel would soon learn when visiting his aunt on his homeward voyage, require the assistance she had been quietly preparing for years. On the summit of Etna, under a clear Mediterranean sky, Herschel began to consider the culminating astronomical endeavor of his career.

HOMEWARD

From Etna Herschel decided to travel back to Palermo through the island's interior, passing through villages that had not seen a foreigner in more than half a decade and horrifying James with meals of boiled snails and sheep intestines. Despite the remoteness of their route, Herschel met another British geological traveler on his way back to Palermo, the Oxford professor and chemist Charles Giles Bridle Daubeny (1795–1867), who had been in Sicily for three months. Daubeny was a younger Oxford colleague of Buckland and was particularly interested in volcanoes, with his own theory of volcanism based on chemical reactions between minerals and subsurface water. Herschel's discussions with Daubeny as they journeyed together no doubt revolved around the chemical analysis of the rock specimens they had collected (and were having hauled for them) across the island. Daubeny certainly shared with Herschel about his earlier studies of the extinct volcanoes of the Auvergne region in southern France. These would soon become the focus of books by Daubeny and Scrope, who both toured the region and arrived at different interpretations of what the extinct volcanoes there indicated about the geological past.[28]

Herschel was anxious to return to the mainland, but unfortunately he and Daubeny arrived back in Palermo at the beginning of the Feast of Saint Rosalia, patron saint of Sicily, to find all commerce in the city shut down for an entire week. The streets were filled with revelers, there were fireworks each night, horse races were held through the main street of the town, and a fleet on the harbor was illuminated with Chinese lanterns in a mock naval battle. The entire thing felt to Herschel like a scene from the *Arabian Nights*. He was exhausted, anxious to depart, and sick to death of cannons being fired throughout the nights; however, even in the midst of the festivities he maintained his consistent observational approach. He calculated the velocity of the racing horses by comparing distances they covered to the beat of his pulse, and when the immense gilded carriage of Saint Rosalia was carried through the streets in procession, he measured its dimensions to the nearest foot (thirty-three feet in height, forty-eight in length).[29]

Finally, the festival ended, and after a month in Sicily, Herschel returned with Daubeny to Naples, departing by the same steam packet on which he had arrived. Daubeny, however, was stricken with malaria as soon as they reached the mainland (Herschel said he had gotten it from sleeping in a marshy place), and Herschel remained in Naples for more than a week to care for him. Malaria was a constant concern for Herschel throughout his travels. He sickened briefly in the city as well, coming down with a fever and growing so weak that at one point he fainted trying to sit up in bed. Within a few days, though, he had recovered and was strong enough to visit the city's observatory in the company of Marguerite Gardiner (1789–1849), the Countess of Blessington,

FIGURE 5.5: Sir John Frederick William Herschel, *Sasso Vernale Valley of Fossa Tyrol from the Gorge of Fredaja*. Graphite drawing made with the aid of a camera lucida, unframed: 8 7/16 × 12 7/8 in. The J. Paul Getty Museum, Los Angeles, Gift of the Graham and Susan Nash Collection.

who had settled in Naples after her own grand tour, and Carlo Brioschi (1782–1833), the observatory's director. As Herschel wrestled with the observatory's large Fraunhofer refracting telescope, which he noted had a horrible mounting that made it difficult to keep pointed at any star, the countess may have shared with Herschel memories of the conversations she had with Lord Byron, whom she had met in Genoa before he left for Greece. (She would publish a book on these conversations in 1834.)[30]

Daubeny took longer to recover, though by the beginning of August he was strong enough that Herschel felt it was safe to leave him behind. Regulations against contagion meant Herschel and James could no longer travel together, as couriers could only take one passenger at a time, so they set out separately. From Naples they retraced their steps northward, traveling by night or in the early morning hours to avoid the heat of high summer. They passed back through Rome and Florence and in Bologna turned eastward, visiting Mantua and Padua before arriving at Venice. Here, Herschel noted the silence of a city without horses or carriages and the hearse-like appearance of the gondolas. Returning through Verona and Padua, Herschel observed with the astronomer Giovanni Sante

Gaspero Santini (1787–1877) at the university of Padua. They measured signals atop Mount Baldo, overlooking Lake Garda, as part of a surveying project by the Austrian government. Herschel had decided to return homeward through the eastern Alps, giving him an opportunity to see the geology of Alpine regions he had not yet explored. At Borghetti, he crossed the Austrian frontier and passed out of Lombardy on his way to Bavaria.[31]

Herschel and Babbage had not ventured this far east on their previous voyage, and Herschel wanted to see as much of the terrain as possible, make more drawings, and collect more minerals to supplement those he had shipped home from Italy. Traveling the length of Lake Garda up the valley of the Adige to Neumarkt, Herschel left James and the coach behind for a ten-day solo expedition into the hills. He traveled light, taking only several burlap collection sacks, hob-nailed boots, a cloak, drawing board, blowpipe, and hammer. Two hours outside of Neumarkt, James caught up with him, carrying Herschel's remaining barometer on his back and hoping Herschel had changed his mind about going on alone. Herschel sent him back with instructions to meet him farther up the road in several days. Staying in village inns and the occasional hayloft, and getting lost at least once, Herschel wandered the Alpine valleys gathering evidence of past volcanic activity. Ten days later, he met up with James in the town of Bozen (today Bolzano).[32]

Crossing the Alps, Herschel arrived in Innsbruck, Austria. In terms of climate and distance, he now felt significantly closer to home, as though he had crossed an ocean rather than a mountain range, and he grew less interested in the landscape he passed through. From Innsbruck, he hurried north to Munich, where he hoped to meet Georg Friedrich von Reichenbach (1771–1826), the instrument maker whose transit circles were in use in almost all the observatories he had visited. He also wanted to meet Joseph von Fraunhofer, whose telescope glass Herschel was realizing was revolutionizing Continental astronomy much as Reichenbach's circles had a generation before. In Munich, however, he found Reichenbach ill and unable to take visitors and realized to his chagrin that Fraunhofer had been in his workshop in Benediktbeuern just as Herschel hurried through the village on his way to Munich. Herschel considered returning, as meeting Fraunhofer while in Germany had been one of his primary objectives, but the glassmaker was supposed to be in Munich any day, and Herschel did not want to risk missing him on the road. Instead, he waited in impatience and uncertainty for two days until Fraunhofer finally arrived.

Herschel's patience was rewarded, and Fraunhofer was happy to show his method and mechanisms for grinding and polishing lenses, though he remained secretive about his method for actually producing the glass. Herschel was especially interested in Fraunhofer's glass as a means of producing the monochromatic light necessary for optical experiments on polarization. High-quality glass that refracted wavelengths of light consistently was also needed for the production of refracting telescopes according to the analytical process Herschel had previously proposed to eliminate aberration. In this respect, and because he was able to make such large lenses, Fraunhofer's glass was revolutionary. It also opened the door for the spectral analysis of sunlight and starlight. By day, Herschel observed the lines that Fraunhofer had discovered in the solar spectrum, and by night, in the observatory of Johann Georg von Soldner (1776–1833), Herschel saw for the first time spectral lines in the stars as well through a prism of four inches in diameter set before the telescope's object glass. Though they would eventually be recognized as the chemical fingerprints of stars, setting the foundation for astrophysics, at the time these lines were seen as a means of measuring the refractive index of glass. Despite these demonstrations, Herschel did not get to see the one thing he most desired: the glassworks themselves, which Fraunhofer kept "enveloped in thick darkness."[33]

The rumors that had reached Herschel in England about Fraunhofer had not been exaggerated. Herschel was impressed by both his products and his methods. "All his process," he noted, were "as far above the blundering, bungling makeshifts of such of our artists as I have seen at work on refractors as his scientific acquirements are superior to theirs." The Bavarian glassworker based his craft on a rigorous, systematic, and experimental foundation that Herschel believed was necessary for the advancement of the field. On the other hand, Fraunhofer kept the details of those methods secret, something opposed to Herschel's own approach to natural philosophy. For all Fraunhofer's success, the secretive nature of his process was still that of the artists' guild rather than the community of the physical sciences. What was needed, Herschel realized, was a public treatment of light that would integrate everything known of both the nature of light and its interaction with glass to provide a summary of the field for experimenters and craftsmen alike.[34]

While in Germany Herschel also met leading German astronomers such as Johann Franz Encke (1791–1865), the most important contributor to celestial mechanics of his generation, and Karl Ludwig Harding (1765–1834), discoverer of the third asteroid, Juno, in 1804. But the most important astronomer Herschel saw was his aunt Caroline, who had moved back to her home country of Hannover. Caroline had been an abiding influence on Herschel's life, and he grew up watching the work she did organizing and finalizing his father's observations. More than simply an observing assistant, Caroline was the means by which William's raw observations became usable scientific data. Though Herschel learned from his father the construction and operation of large telescopes, it was from his aunt that he learned to reduce and organize astronomical observations and coordinate a systematic observing project. At the time of William's death, Caroline had been living in England for fifty years and had gained international renown for her own work, specifically her discovery of comets.

After the death of her brother, Caroline returned to Hannover, where she still had siblings and an extended family. (Though a German state, Hannover had been governed by the British royal family since 1714, when its ruler ascended the British throne as King George I.) In the years since, she had worked consistently on a master list that organized all of William's nebulae, stitching together observations that had been published in multiple catalogs and presenting them in a single, unified list in order of their positions in the sky. This resource was a vital step in resuming the nebula surveys where William had left off, something Caroline was anxious for her nephew to begin. When Herschel arrived in Hannover to see his aunt, they discussed the work that Caroline was still completing, and Herschel urged her to published it. Instead, she entrusted it to him, telling him she had created it for his use alone. The immense list, which Caroline sent him when completed, became the germ of Herschel's subsequent observing project. Herschel had started his astronomical work revisiting his father's double star observations in search of possible changes in those objects. Caroline's catalog allowed him to do the same for the rest of the sidereal heavens and perhaps answer the question that had haunted his father for decades: whether it was possible to observe change in the nebulae themselves. If the skies of Sicily provided a glimpse of what extending his father's work might entail, Caroline's catalog provided the foundation to actually begin the task.

In Hannover Herschel was reacquainted with his father's family, meeting uncles, aunts, and cousins, many for the first time. The house where his father was born no longer stood, but he was able to give his mother a good account of Caroline and the rest of his family. The final leg of Herschel's journey was from Hannover to Rotterdam, where he disposed of the carriage that had carried him so far across Europe, and made the rest of the journey by steamship via canal from Rotterdam to Amsterdam and thence on to London, avoiding overland travel from Dover. After hundreds of miles across France and Germany and Italy, Herschel was quite finished with land travel and anxious to return home. He had no intention of waiting for favorable sailing conditions to cross the Channel. It was now steam all the way: "*Steam* is a *sine qua non* with me for all water travel." By the end of October, after a grand tour of nearly seven months, Herschel was home.[35]

Herschel's grand tour had taken him from London to the southernmost reaches of Europe. Along the way, he had cast his international net of collaboration even farther afield, and the new connections he made would become a consistent part of his correspondence for the rest of his career. Besides his colleagues in Paris and astronomers in Germany, Herschel had visited nearly every working observatory in Italy, in each carefully examining their instruments and the glass of their telescopes. This aspect of his tour culminated with viewing Fraunhofer's glass and the realization that the Bavarian glassmaker had accomplished what Herschel had long been advocating: establishing glassmaking on an experimental footing. Immediately upon his return to Slough, Herschel began his answer to Fraunhofer, a work that would put the current state of optics on display for all.

Herschel had come a long way from his days of mathematical analysis. His own work was now reaching for a wider and broader scope. The Herschel who recorded everything on his trip throughout Europe, from weather patterns to geological strata to the physical dimensions of Greek ruins, was straining to pull together all he saw into a cohesive interpretive framework. That framework, however, would not be purely abstract. Herschel's emerging vision of a new form of natural philosophy was becoming married to—not in tension with—a Romantic vision of the world. In the Alps, Herschel learned that mountains were laboratories of atmospheric and geological physics. Atop Vesuvius and Etna, he saw the chemical furnaces of nature at work on a vast scale. In the ruins of southern Italy, he perceived clues of the slow geological sweep of time and change. And finally, under Sicilian skies, Herschel glimpsed the riches the southern heavens offered. Herschel was realizing the laboratory of the world needed more than mathematics to interpret. It needed an imagination shaped by scientific virtues—virtues that, Herschel would soon have reason to believe, required renewal and reform among the scientific elite of London.

CHAPTER 6

A LONDON DISCONTENT

I am perfectly sick of the life of a savant by profession which leads
to nothing but quarrels & misunderstandings in which every one's
temper is soured and no one's real interests are advanced.

—JOHN HERSCHEL TO CHARLES BABBAGE, FEBRUARY 1828

By the conclusion of his European grand tour, John Herschel's prestige
as a natural philosopher had only grown. To the credibility established
by his early mathematical work he had added a reputation as an expert
experimentalist and an authority on optical and chemical research. His
trips abroad strengthened his ties with natural philosophers throughout
Europe, many of whom viewed Herschel as their primary point of contact with the British scientific community. Upon his return, Herschel's
unique instrumental and observational resources, the inheritance of his
father and aunt, ensured him an avenue for continued discovery and
publication. Institutionally, his roles in the Astronomical Society and
Royal Society meant there was little in natural science that he did not
partake in or discussion he was not privy to. In sum, by 1824 Herschel
had established himself as a center, if not *the* center, of gravity in the
British scientific community.

Scientific influence, however, was mediated through institutions,
and the primary institution for the pursuit of natural philosophy in Britain remained the Royal Society of London. Like Isaac Newton before
him, Herschel embodied one of the poles the Royal Society had his-
torically balanced between. Herschel was not, as many of its members
remained, an aristocratic enthusiast who viewed natural philosophy as a

hobby or pastime, one of those whose numbers both supported the society financially and brought it social and political prestige. Rather, Herschel was of the much smaller membership of fellows pursuing active research who published in the Society's *Philosophical Transactions* and who viewed the society as a tool for the progress of the sciences rather than for personal social ends. Maintaining a balance between the desires of active members like Herschel and the patronage, support, and influence of the society's aristocratic fellows was a delicate task, and when Herschel returned from abroad there was no one who knew this better than Davies Gilbert (1767–1839).

An influential member of Parliament from Cornwall, Gilbert had himself been a natural philosopher before entering politics. As part of Joseph Banks's ruling circle in the society, he had been instrumental in connecting Humphry Davy with the opportunities that had led the young chemist to fame and his eventual presidency of the society. In the House of Commons, Gilbert was a primary government supporter of science and was instrumental in securing the unprecedented funding provided for Charles Babbage's calculating machine, invigorating the Board of Longitude, and initiating construction of the observatory at the Cape of Good Hope in South Africa, a southern counterpart to Greenwich. For Gilbert, maintaining the prestige and influence of the Royal Society was essential to making such projects reality.

Gilbert, however, was of the previous generation when it came to managing scientific endeavors and governing the society, and his conservative position soon brought him into conflict with Herschel and other reformers who believed the Royal Society needed an organizational overhaul. Initially a proponent of parliamentary reform, Gilbert, like many conservative politicians of the time, became increasingly suspicious of democratic upheaval in the wake of the French Terror and the upwelling of radical populist movements throughout Britain. In his political career, he walked a tightrope between supporting modest reform to limit privilege and corruption in the government and opposing the more radical reform he believed would lead to mob rule and the overthrow of institutions. Gilbert carried this stance into governing the Royal Society as well, resisting attempts to weaken the privileges of the society's president and aristocratic elements or strengthen the role of the active, more scientific members. For the prestige of the Royal Society and its continued relationship with government, the reforming spirit threatening the country needed to be prevented from tainting the society.[1]

At the time of Banks's death, when Herschel hoped autocratic rule of the society was at an end, Gilbert was a member of the society's governing council and a vice president. With Banks gone, Gilbert found his conservative positions in immediate tension with demands of reformers like Herschel and worked to have Davy elected as the next president. Davy, recognized for his scientific accomplishments and from a modest, non-aristocratic background, seemed to Gilbert an ideal compromise between the aristocratic and reform-minded parties within the society. When Davy became president despite Herschel's opposition, Herschel was nonetheless named one of the society's two secretaries, likely due to Gilbert's policy of compromise and reconciliation. Gilbert had no desire to change the governing structure of the Royal Society, but he recognized the importance of practitioners like Herschel for the society's legitimacy.

Herschel had reached a position where his views on natural philosophy and its practice had significant influence. Beside his role as Royal Society secretary, Herschel was soon nominated president of the Astronomical Society. He continued to serve as commissioner to the Board of Longitude and worked alongside Michael Faraday and George Dollond on the Joint Glass Committee organized by the Royal Society and the Board of Longitude, where he had responsibility for testing the manufactured glass it was hoped could compete with what he had seen in Europe. With these roles, the period immediately following Herschel's return from Europe could have been the culmination of the reforming project he had begun in Cambridge so many years before. But rather than the freedom to reinvigorate British natural philosophy, as he hoped, Herschel grew steadily disillusioned by the responsibilities, debates, and disagreements *about* science that he believed continually got in the way of the effective practice *of* science. He found a dismaying amount of contention and controversy at the heart of the scientific community and became convinced that the British government neither respected nor supported sciences, even as he felt his country continued to fall behind the Continent in research. When, after years of friction, the tensions between Gilbert and the other reformers erupted into acrimonious controversy, Herschel would abandon the institutional positions he had worked so hard to obtain and retreat from London to craft his own appeal for a new ideal of the scientific life.

"LIGHT"

Herschel was elected one of the Royal Society's two secretaries at the meeting of 25 November 1824, soon after his return from his extended European tour. The role of secretary required significant involvement in the day-to-day administration of the society, including regular attendance at all meetings, in addition to the council meetings where much of the business of the society was handled. Herschel recognized his responsibilities would make it nearly impossible to continue splitting his time between Slough and London, and he once again took permanent lodgings in the city.

At the same time as his institutional responsibilities increased, Herschel began a major new project prompted by his recent travels. His earlier experimental work had focused on optics, especially on questions of the nature of light and polarization, and his European tour showed him the enduring contentions among natural philosophers in this field as well as its importance for continued development of quality glass. Different experimenters in different national contexts pursued the topic without a complete grasp of what had been done before or the experimental results others had achieved. Herschel realized that a comprehensive, synthetic treatment of light and optics was needed to summarize the state of the field, guide research, stimulate further development of optical glass, and—hopefully—extinguish the priority disputes that continued to smolder. Along the way to providing this, Herschel would address the two competing theories of light and develop his own views of the nature of natural philosophy—which was coming more and more to resemble modern science.

This opportunity came through an invitation to write an article on light for the *Encyclopedia Metropolitana*. Early nineteenth-century encyclopedias, marketed toward a growing middle class and literary public, provided extensive and sometimes surprisingly technical overviews of various subjects, and Herschel would eventually contribute highly mathematical articles on astronomy and sound to the *Metropolitana* as well. He jumped at the chance to author a long article about light, as preparing it would give him an excuse to read widely in the field and stay acquainted with the latest experimental results. "I am anxious in everything I write," he told his editor, "to *learn* at the same time that I teach." Writing the article gave Herschel another opportunity to connect mathematical analysis with the physical world, but it also gave him a platform to argue for his vision of science. In short, "Light" would provide a test case for articulating the scientific virtues Herschel was continuing to develop.[2]

Herschel began his account, which ultimately ran to more than two hundred pages, with an appreciation of sight, which he described as the most perfect and accurate sense. The beauty of optical phenomenon, which he had spent so many hours studying in his darkened rooms at Slough or alongside colleagues in Paris, left him "lost in amazement and gratitude." For Herschel, grown up in the shadow of huge telescopes, these instruments were significant not only for the objects viewed *through* them but also for the light captured *by* them. Optics connected his experiments on the properties of glass with his prior work on the sensitivity of different chemicals to light, work that had already set the groundwork for early photography, though that particular application would not bear fruit for many years. Yet the interactions between light and glass or light and chemicals were only illustrations of a deeper truth: light was fundamentally a tool to investigate nature, "to feel the ultimate molecules of natural bodies."[3]

Optical laws were mathematical laws given substance. Not even physical astronomy, often seen during this period as the epitome of mathematics applied to the universe, could make this connection as clearly and straightforwardly. The solar system was a complex physical system with so many variables that astronomers were forced to make choices about what to approximate in their equations and what to simply ignore. Light, in contrast, traced out pure mathematical expression. Reflection and refraction, for instance, took place "in exact conformity with the results deduced from them by mathematical reasoning." With light, Herschel saw analysis most clearly touching the physical world, and the mathematics he employed to represent all aspects of known optical phenomena, from reflection and refraction to diffraction and polarization, encompassed the tools and methods he had advocated since Cambridge. This was the final triumph of the Analytical Society: that the most up-to-date and complete accounting of the most advanced experimental science was presented—and could *only* be accurately presented—through the language of analysis.[4]

Central to Herschel's account was the discovery and explanation of the polarized nature of light, an example of successful research that Herschel considered to be the most important development in natural philosophy since Newton's gravitational synthesis. Experiments on this topic, which Herschel had spent much of his early career pursuing, had opened up a field "so singular and various" that it was "like entering into a new world—so splendid as to render it one of the most delightful branches of experimental inquiry." Yet as Herschel had witnessed firsthand, the scientific triumph of polarization was tangled in bitter disputes about who had discovered what aspects of the phenomenon first. Herschel purposefully avoided attributing credit to any of his colleagues. He did not, he told a friend, want to get involved with those who "quarreled ad nauseum about their claims and their priorities and their discoveries" and attacked anyone "fool enough to meddle with their bone of contention." Herschel had worked alongside most of those involved, conducting experiments with Jean-Baptiste Biot in Paris and sharing and comparing results with David Brewster in Edinburgh. If light was the manifestation of pure mathematics, the field of optics was the opposite: a chorus of bickering that deflected energy from its pursuit and appreciation. Herschel was explicit in "Light" about disputes "carried on in a spirit of rivalry or nationality." They were "utterly derogatory to the interests and dignity of science, and . . . little short, indeed, of sacrilegious profanation of regions which we have always been accustomed to regard as a delightful and honorable refuge from the miserable turmoils and contentions of interested life." "Light" made crystal clear Herschel's ideal of natural philosophy devoid of contention and personal strife.[5]

Part of the problem was poor communication between the Continent and Britain (a disconnect that Herschel's repeated trips to France had attempted to address), which meant results were not effectively shared and researchers in one country could be unaware of what had already been done elsewhere. Herschel hoped the synthesis and overview that "Light" provided would fix this, and, indeed, soon after its publication the book-length article circulated widely and appeared in both French and German translations. The "bewildering detail" of all observational facts relating to polarized light were still being "merged in that of the laws from which they flow." In other words, there was not yet an adequate comprehensive theory of light. For laws to be understood, or even adequately described, they had to be interpreted in a theoretical framework. But neither the undulatory or corpuscular theory of light—neither describing light as a wave or as particles—was yet sufficient: neither furnished "that complete and satisfactory explanation of *all* the phenomena of light."[6]

Composing "Light" allowed Herschel to celebrate the experimental and theoretical progress of the field, advocate how the science should best be pursued, communicate what had been done so far and what questions remained to be answered, and exemplify the unity between mathematics and nature. In sum, "Light" presented Herschel's most complete vision thus far of what an effective science was and how it should work. Yet even as Herschel completed this treatment of his scientific ideal, the "turmoils and contentions of interested life" challenged his conception.

"A DEAD LETTER": REFORMING THE ROYAL SOCIETY

As Herschel composed "Light," his own experimental work in optics had largely ceased apart from his work with the Glass Committee. The plan for this committee had been for Faraday and Dollond to create glass samples and for Herschel to test their optical properties. Together, the three hoped to develop methods that would allow Britain to compete with the exceptionally clear glass coming out of Joseph Fraunhofer's workshop in Benediktbeuern. At the same time, in the summer of 1825, Herschel was pursuing a project to literally connect the French and British scientific communities. With the astronomer and army officer

Edward Sabine (1788–1883), Herschel coordinated with French astronomers to determine the precise difference in longitude between the observatories of Paris and Greenwich, a project he had advocated since he visited Paris in 1819.

Throughout that summer, Herschel and Sabine, who had previously traveled on Arctic expeditions with the navy and worked to calculate the shape of Earth, led teams of artillery officers firing rockets from a series of observation stations linking the two observatories across the Channel and miles of countryside. By measuring when rocket flashes were seen at the different stations, they hoped to determine the precise distances between them. Sabine coordinated the experiment on the French side of the Channel, while Herschel had charge of soldiers on the English side. Though the operation was unsuccessful (many of the rockets did not fire correctly or were not high enough to be seen), the project gave Herschel the chance to put his view of international scientific cooperation into effect.

Herschel also remained active on the Board of Longitude, where his vision of natural philosophy occasionally put him at odds with other members. Ostensibly, the board was authorized to support methods of determinations of longitude for navigation. Yet for a natural philosopher like Herschel, every expedition was an opportunity for discovery, and he pushed the board to give naval officers greater leeway for more data to be gathered by ships on their voyages. Often this meant the board advised on what kinds of scientific instruments British naval vessels should carry and provided the instruments on loan. Even this approach could run into resistance, though, as in the fall of 1825 when Herschel was warned by Thomas Young, board secretary, against arguing that naval officers be supplied camera lucidae, sextants, and barometers—the same instruments Herschel had carried with him on his European travels. There was no need, Young and the Admiralty argued, for such distractions. Herschel believed naval vessels should be used as platforms for an increasingly wider collection of data from around the globe, but others on the board and in the government saw this as a needless expense, complication, and distraction.[7]

Finally, Herschel's work with the Royal Society forced him to navigate a Humphry Davy presidency. Herschel's initial doubts regarding Davy's fitness for the position were quickly confirmed. To Herschel's dismay, Davy soon filled society meetings with his personal quarrels and controversies, particularly against David Brewster in Scotland. Herschel recorded his frustrations in his journal—as well as a resolution to keep from being drawn into them, perhaps remembering the way he had allowed South to involve him in embarrassing arguments in the past.

"Take care not to mix in it," he noted to himself at the end of 1825 as a "1st principle." If forced to have an opinion on anything, "hear both sides fully." Herschel admitted that the disagreements Davy was involving the Royal Society in confirmed the predictions he had made "when urging Wollaston to stand for PRS." Davy's personality remained a liability to the pursuit of science within the society.[8]

These frustrations, however, were simply tremors preceding a larger controversy that broke out late in the following year, this time centered around Charles Babbage. The position of second Royal Society secretary was coming open, and Herschel hoped Babbage would get it. Babbage had become a public figure due to his government grant for the construction of a calculating machine that promised to make the complex calculations needed for everything from stellar positions to life insurance annuities as mechanized as a steam engine, but he held no official scientific roles beyond membership in societies like the Royal Society and the Astronomical Society. Being elected a secretary to the Royal Society would provide an important confirmation of his work and a further triumph for the reforming party. Having Herschel and Babbage, both recognized scientific practitioners, as the society's two secretaries would increase their influence and perhaps allow them to temper Davy's destabilizing role.

The plan was for James South to nominate Babbage for the position. South was an active reforming voice in the society as well, though his fiery temper and occasional lapses of judgment had already created headaches for Herschel. Herschel, who no doubt felt as secretary it would be improper to nominate Babbage himself, nonetheless wrote a strong letter to Babbage urging him to go along with their plan and acquiesce to the nomination. He stressed Babbage's support among the members of the society known for active scientific work: "[Henry] Fitton is ardent . . . [Francis] Beaufort is decisive, Baily is sure, [Henry] Kater is enthusiastic, [Thomas] Colby is keen, Gilbert will vote, [Charles] Lyell jumped & rubbed his hand with delight at the thought." In short, Herschel concluded, all their reforming allies were "anxious to do what they do best in the best way to secure your services" to the Royal Society.[9]

There was at least one other person being considered for the position, the chemist and zoologist John George Children (1777–1852), but Herschel assured Babbage that Davy planned to leave the decision up to the vote of the Royal Society Council. Because so many of Babbage's friends and supporters served on the council, Herschel looked on Babbage's election as a "sure thing." Herschel was already planning to step down when his own term of secretary ran out the following year, but he told Babbage that serving with him in the time that remained would be "one of the few causes which will make me regret its termination." In the year they would have together, Herschel was sure they could accomplish much toward making the society more democratic and supportive of its scientific members. Convinced there would be no serious opposition, Herschel and colleagues proceeded with their plan to nominate Babbage at the next council meeting.[10]

Their plan, however, foundered on Davy's mercurial disposition. On Thursday, 23 November, Davy summoned a handful of council members, including Herschel, to a special meeting ostensibly to address some outstanding financial business. At the meeting, which seemed arranged to circumvent the support Herschel and his allies had been building for Babbage among the wider council, Davy surprised everyone by reversing his decision and supporting Children for the secretaryship. Babbage had already been nominated, but Davy's about-face negated Herschel's assurance to Babbage that his election would be unopposed. Without the president's support, at worst Babbage would lose and at best he would win in a contested election that saw him pitted against Davy's favored candidate. Under these conditions, Herschel felt compelled to withdraw Babbage's name. Babbage, he made clear, desired to serve only with "*a full and fair recognition of his merits in open Council.*" Wollaston, present at the meeting and as surprised and disappointed as Herschel, asked Davy whether he meant to use the privilege of his office to appoint the secretary himself or open it to the wider council for a vote. Davy made it clear there would be no vote, and there, to Herschel's immense frustration, the matter rested.

Herschel left the meeting furious and disillusioned, and not simply because the outcome meant Babbage had lost the position of secretary. In Herschel's mind, Davy's behavior was a perfect example of the enduring cronyism and favoritism at the core of the Royal Society. Banks's presidency had been characterized by positions handed out to friends or supporters at the president's whim. Herschel, despite his doubts about Davy, had hoped this aspect of society governance had changed. That clearly was not the case, and Herschel resolved, after his term of secretary was over, to serve "no longer under *any circumstances.*" He was fed up with politics and favoritism. Only the nearness of the end of his own term as secretary and the "disgraceful broil" which a resignation under the circumstances would create kept Herschel from making a statement by doing so. It was now clear to him, however, that the Royal Society's issues would not be addressed under Davy's leadership.[11]

If there was a bright side to this disappointment, Herschel tried to convince himself and Babbage, it was that Davy's actions made the problems within the society clear. "I am anxious at present not to think of the past with annoyance, but to the future with hope," he consoled Babbage. Davy's behavior had created among Babbage's friends "a feeling which though at present rather somber is rapidly changing into a resolve to place the future conduct of the society on a higher and better footing." At the same time, Herschel knew Babbage's own reaction could undermine this "higher and better footing." As he wrote to Baily, one of Babbage's supporters who had been excluded from the meeting, the best course was one "of deliberation, composure, and dignity." The key for Babbage's party, which, Herschel claimed, embraced "nearly all the really scientific part of the body," was to continue patient reform from within and not resort to public controversy that would undermine the society's influence. Herschel believed reserve was the best course of action in response to Davy, but he feared such a course would be difficult for Babbage—and, unfortunately, he was right.[12]

For Herschel, Davy's behavior showed that the autocratic style of running the society was alive and well. For Babbage, being deprived of the position of secretary was a match that lit the fuse of his explosive personality. That fuse might simply have smoldered for a time, but less than a month after this disappointment Davy added insult to injury when Babbage's certificate to elect a colleague for membership in the society was rejected on a technicality. Babbage was furious that Davy would be a stickler in this case when he bent the rules to suit his own intentions elsewhere. Finally, Babbage was additionally incensed when he was passed over for one of the society's new medals. He felt the society had again violated its own rules by giving the award to work published outside of the dates originally specified for work to be considered. He prepared to take his grievances against the Royal Society and its president public.[13]

Herschel's concerns were confirmed the next month when he received a letter from Georgiana, Babbage's wife, warning Herschel that friends had come to see Babbage and were planning to make difficulties at the next society meeting—exactly the kind of controversy Herschel hoped to avoid. Georgiana urged Herschel to skip the meeting, using his mother's illness as an excuse. "Charles often says," she told Herschel, "'the only restraint I feel is the fear of annoying Herschel.'" Instead of convincing him to stay away, this confirmed Herschel's need to intervene and keep things from spiraling out of control.[14]

Herschel's first step was writing Babbage directly. It was normal to be disappointed, Herschel commiserated, first at the authoritarian appointment of the new secretary and then by the pedantic dismissal of Babbage's membership proposal, but science itself would suffer if the Royal Society was riven by petty bickering. Herschel urged Babbage to leave things alone until they blew over, arguing that he was hurting his own case. "You have adopted of late a tone of acrimony & severity . . . which I am certain you will hereafter regret, because it is indiscriminating, and wounds alike your friends and those who are indifferent to you." The only way forward, Herschel argued, was to avoid discussing the issues, public or private, "relative to the management of the affairs of the Society, right or wrong." Let the offense go. "For God's sake dismiss it from your mind."

There was more at play here than Herschel's personal tendency to avoid dispute. The status of the Royal Society, Herschel felt, was tantamount to the dignity and the influence of natural philosophy in Britain, something Herschel already feared was waning. "The credit & dignity of the Society will be sacrificed by any further agitation of the point in public under the present state of feeling among its members," Herschel maintained. Nothing should happen at the next meeting "to compromise the outward decorum of its proceedings." As had been the case with their mathematical reform, Herschel still believed change in the state and status of science could only come about within the institutions that already existed. "It is not the course to begin reform," Herschel concluded, "by the degradation of the body to be reformed." Contention within the Royal Society would expose it to public discredit and in the long run hurt British science itself.[15]

Herschel's frustrations with the Royal Society coupled with his insistence that any reform be measured and conducted within the bounds of the very institutions to be reformed mirrored what was happening in the broader British political world. In Parliament politicians were trying to balance growing calls for reform—specifically to laws on agricultural imports and prices, the complete abolition of slavery, and allowing nonconforming Christians (particularly Catholics) to hold public office—without giving way to the broad democratic fervor of the French Revolution and its aftermath. Most acknowledged that the kingdom's representational structures were outdated (many booming new factory towns, for instance, had no representation in Parliament at all, while other districts known as "pocket boroughs" or "rotten boroughs" were controlled by landholding aristocrats) but the majority of reforming parliamentarians, mainly Whigs, wanted the same kind of measured, gradual reform Herschel was advocating in the Royal Society.

For years now, Herschel had been fighting what felt to him an uphill battle for the improvement of natural philosophy—which was, throughout Herschel's career, becoming more and more identified with the physical and mathematical sciences—from his student days when he and Babbage had attempted almost singlehandedly to transform British mathematics, to his efforts now to produce glass that would protect British prominence in instruments, to his urgings for scientific expeditions in the face of a government that seemed indifferent to the reach and resources of the British navy in the service of science. Herschel was reminded of this struggle as he wrote "Light," finding that almost every citation he provided for advancements in the theoretical and experimental understanding of light came from journals or publications outside of Britain. Despite his best efforts, Herschel confided to his aunt Caroline, he felt that in England science was "going to sleep." He had slowly and steadily built the credibility to take meaningful steps to improve this situation, steps that were more important now than ever, but all this might fall apart if Babbage insisted on rancorous public controversy.[16]

Herschel had reason to fear. Babbage's anger was great enough that even Herschel's friend Grahame, who moved outside scientific circles, took note: "God grant Babbage more dignity of disposition and strength of mind." Herschel told Grahame he had succeeded in restraining Babbage from "taking a public step which would have had the effect of setting the whole Royal Society together by the ears" but that in so doing he may have exhausted his personal credibility with their mutual friend. "The whole of this affair has mortified me beyond any thing," he confided, worn out with the issue and the personal dramas it entailed.[17]

Despite the dangers posed by Babbage's personality, the controversy over his election to secretary did indeed catalyze the beginnings of some of the reforming movements Herschel wanted to see. At a subsequent Royal Society meeting, Babbage brought up an earlier failed proposal to address the process of admitting new members, a question that had become a central issue for those urging reform. The majority of Royal Society members were not active in any type of scientific work, and the value of the title of fellow, many felt, had become devalued to little more than a symbol of social status. Herschel's complaint to William Whewell about all those who "let their FRS [fellow of the Royal Society] remain a dead letter" was by this time a common one. Babbage's motion prompted an analysis of membership, which showed to no one's surprise that the vast majority of the research and scholarly activity of the society was conducted by a small percentage of members. A committee was set up to consider revisions to the process of admitting new fellows, with an eye to emphasizing scientific achievement as a requirement for membership. To the dismay of conservatives like Gilbert, this committee consisted entirely of core reformers in the society, including Herschel, South, and Babbage.[18]

At the final society meeting before the 1827 summer recess and before Herschel was set to step down from his duties as secretary, the membership committee made its formal recommendations, proposing a membership cap of four hundred for the society and the admission of only four fellows per year until that number was reached. The committee also recommended that leadership positions be nominated by the outgoing council, removing power from the president and putting it into the hands of the senior fellows. Gilbert, as vice president, signed off on these recommendations, even though they represented the kind of broader control he opposed and believed would undermine the society's influence.[19]

With the committee's report, Herschel was satisfied that modest reform had been initiated, and he formally announced his resignation as secretary. He had already expressed his intentions to leave the post in order to devote more time to astronomy, but brewing tensions with colleagues like Babbage made him all the more eager to relinquish the role. In his resignation letter to colleagues, Herschel confessed that fulfilling the duties of secretary had become "incompatible with my other pursuits," namely his observing projects in Slough. The version of his resignation letter that Herschel sent to Gilbert, however, was more pointed. Herschel told Gilbert he would have had more regret resigning as secretary "had it not been recently pressed on my attention with somewhat of a painful distinctness," referring to Davy's handling of the secretaryship, that the position was still regarded as the president's prerogative. Herschel emphasized that when he had taken the position, he had not seen it as such but instead considered it an elected position previously filled by philosophers of distinction like Robert Hooke and William Wollaston. Davy's actions had undermined the nature of the role of secretary to the society, making it, Herschel felt, "a matter of mere routine and clerkship" instead of "an important scientific trust."[20]

His hopes for the society reduced to modest reform of membership policies, Herschel prepared to withdraw from London and devote his time more exclusively to astronomy. But was astronomy drawing Herschel from London, or was London pushing Herschel into astronomy? Herschel's roles and responsibilities at the center of scientific society, it seemed, had become incompatible with the actual practice of natural philosophy. Certainly, Herschel needed the clear skies of Slough for his sweeps for nebulae. But his retreat from the Royal Society revealed what Herschel had learned of the business of science in London: that it was rife with contention and politics and had little to do with what he perceived as actual scientific endeavor. In as much as "Light" presented light as the embodiment of mathematical analysis, so scientific practice, Herschel was coming to believe, might best be embodied by a moral character working independent from the social circles of natural philosophers.

A SCIENTIFIC RETREAT

William Herschel's nebulae project remained a great uncompleted frontier, and Caroline's work collecting and organizing William's thousands of nebulae from his multiple catalogs into a single resource provided the catalyst for the younger Herschel to resume this project. Caroline and Herschel had discussed the project when he visited her at the conclusion of his European grand tour, and by March of 1825 the revised catalog was in his hands. The catalog, along with the refurbished twenty-foot reflector, gave Herschel the unique tools necessary to resume and complete William's surveys of the sidereal heavens. Herschel's double star work had helped spur the work of other observers, like Giovanni Amici in Italy and Friedrich Georg Wilhelm von Struve (1793–1864) at the Dorpat Observatory in the Russian empire (modern-day Estonia), but no one apart from Herschel yet had the capacity to reobserve the nebulae and star clusters that his father had discovered. Those objects would become, Herschel assured his aunt, "my especial charge," as "nobody else can see them." Since his return from Europe, Herschel had attempted occasional observations from Slough, but his responsibilities in London had made this nearly impossible. He had "so much upon my hands," Herschel admitted to Caroline, that he was "in a continual fever of the spirits."[21]

Nebulae and star clusters, like double stars, were targets that pushed astronomy beyond its traditional bounds of measuring motions and positions within the solar system. In particular, observing nebulae offered a chance to address questions of the evolution of stellar systems. William Herschel's discovery of 2,500 new nebulae and star clusters had led him to speculate on their nature, and the key to answering whether these objects were stars in the process of forming or simply distant systems of discrete stars was to observe them over time and determine whether they changed. Well-known nebulae, such as the Andromeda Nebula and the Orion Nebula, were targets observers often claimed to have witnessed changing in structure over time, but John Herschel doubted these reports and recognized that only careful observations of large numbers of nebulae could decide the question. As with double stars, the dynamic nature of the sidereal heavens would be revealed by careful and systematic surveys. It was this immense project for which Herschel was now uniquely equipped.

Sweeping for nebulae required time, patience, and clear, dark skies. The method Herschel's father had developed, based on Caroline's early observing practice, was to hold the large twenty-foot telescope pointed in a specific direction in the sky and allow Earth's rotation to carry objects through the telescope's field of view. By moving the telescope slightly up and down, a slice of sky could be "swept out" each night, revealing the nebulae, star clusters, and double stars it held. Herschel's goal was to sweep areas of the sky his father had already surveyed to confirm William's discoveries. He would determine the position of each nebula with a higher degree of precision than his father's catalogs had provided and describe the appearance of each to establish a baseline for subsequent observers and reveal whether any drastic changes in form had taken place since William's original observations. This project, conducted exclusively from Slough, would ultimately last several years, comprise nearly 430 sweeps, and result in a catalog of more than 2,300 objects, published in 1833.

Herschel took his first tentative steps in this immense project in 1825, upon his return from his European tour, recording a single sweep one month after receiving his aunt's catalog and another sweep later that year, but his responsibilities in London meant the project proceeded slowly, as trips to Slough for observing were time-consuming and tedious. On top of the occasional observing nights, Herschel was also still hard at work composing "Light." In 1826 Herschel published an overview and introduction of his planned survey, outlining his progress so far. The paper included another catalog of a few hundred double stars, an overview of his method of sweeping and the structure and design of the twenty-foot telescope, and a series of observations of both the Orion and Andromeda Nebulae. Herschel outlined the primary challenges of nebular observations: precise measurements of fuzzy objects that could vary in appearance based on weather conditions, quality of instrument, and skill of observer. The nebulae, he made clear, would tax his skills as a meticulous and systematic observer to the utmost.[22]

By early 1827 both frustrations with the London scientific scene and the growing realization of the time the nebulae would require led Herschel to retreat completely from the city. Stepping away from responsibilities, however, was complicated that year by his election to the presidency of the Astronomical Society. This position was another opportunity to lend his credibility and leadership to reform, heading the society he had helped create almost a decade earlier. But by this point, Herschel had learned the limits in effecting change through institutions. To his chagrin, the Royal Society's controversies were beginning to spill into the Astronomical Society as well. At a meeting at the end of 1827, for instance, Herschel had invited the amateur astronomer and naval officer William Henry Smyth (1788–1865) to speak, only to have him attacked by South so violently Herschel felt obliged to write Smyth an apology, assuring Smyth that the society supported and appreciated his work despite South's "not a little provoking" attitude. Serene nebulae and long nights of observation were contrasting more and more with evenings spent in the city arguing about astronomy instead of practicing it.[23]

By 1828 Herschel wanted to be done with the Astronomical Society as much as he wanted to be free of the Royal Society, and his nebulae work offered the perfect excuse. Herschel needed dark skies to spot nebulae, and there were only so many nights when this was possible. Nearly every meeting in the coming year, Herschel explained to Francis Baily in a letter informing his friend of his intention to resign from the presidency, happened to fall when the moon's phase made observing ideal. "A *pellucid* sky, the *total* absence of moonshine and twilight, and *nebulae to look at* are conditions which coincide on the average not 20 nights in the year," he explained, and the sacrifice of a single one was precious. By now, Herschel was pursuing the nebulae in earnest. Making his father's work useful to a new generation of observers had become "a sacred duty which I cannot postpone to any consideration."[24]

Herschel also added an additional telescope to use alongside the twenty-foot reflector, a seven-foot equatorial he purchased from South with which he had begun his original double star project. Herschel hoped to use this telescope to search for parallax among double stars according to the method he had outlined in a *Philosophical Transactions* paper two years before. Herschel kept his London friends updated on his observations, but his focus shifted away from participation in the institutions of science to the pursuit of his own projects. His social life turned inward as well, and he began declining invitations in order to ensure as many nights as possible for observation. Pursuing astronomy at Slough, he was "heartily glad to be out of the cabals and contentions of the *learned*." Since Cambridge, Herschel had worked to gain access to the inner circles of scientific influence. Now he was disengaging, convinced his time was better spent pursuing his own work on the scientific periphery. Even in this withdrawal, though, the aftershocks of failed reform and ongoing contention would reach him.[25]

FAILURE OF REFORM

Gilbert could not have taken Herschel's rebuke at resigning the secretaryship too much to heart, because at the time of Herschel's resignation he was in the process of doing the very thing Herschel condemned. For months Gilbert had been in talks with Robert Peel (1788–1850), a conservative member of Parliament who most recently had served as home secretary and would eventually go on to two terms as prime minister. Gilbert hoped to convince Peel to become Royal Society president upon

Davy's impending resignation due to the latter's poor health. In Gilbert's eyes, Peel was the ideal candidate, an example of the kind of gentlemanly leadership the society needed. Davy, Gilbert believed, had shown the error of having a practicing natural philosopher as a president instead of choosing a gentleman who happened to support science. People like Herschel felt Davy was a poor president because he had stopped behaving like a philosopher and had instead become arrogant and ridiculous upon his elevation to the presidency. Gilbert, on the other hand, saw Davy as an example of the type of president that resulted from someone without the social training of gentility. For Gilbert, Davy simply did not have the proper decorum and demeanor.

Peel was Gilbert's answer to the dilemma Davy's presidency had created as well as the Royal Society's protection against reform. A "Great Contest," the conservative member of Parliament told Peel, was playing out in the society parallel to that in the country itself. The society was being riven with the same "conflict of Aristocratic and Democratic Opinion" that Gilbert saw spreading across the countryside. "I wish the Royal Society rescued from the latter," he told Peel. Gilbert was a Tory, a term just coming into common usage, and viewed with great suspicion the reforms Whigs were urging in politics, believing they would result in the same democratic chaos he feared would engulf the Royal Society. Unfortunately, like Babbage, who had refused a non-unanimous election to the position of secretary, Peel was not interested in wading into a situation where there was dispute about whether he was wanted for the job. Peel would only stand for election to the presidency, he told Gilbert, if his nomination was uncontested.[26]

Gilbert's chance to make Peel president (using exactly the sort of backroom dealing Herschel opposed on principle) came when Davy finally announced his resignation in July 1827, only a few days after Herschel resigned as secretary. The aging chemist's resignation was not unexpected, as he had already left London for Europe due to health issues. Davy would never return to England; he died in Geneva two years later. But now the path was opened for a Peel presidency that Gilbert hoped would steer the society back toward the dignity and influence it had held under Banks. During the summer recess, Gilbert tried to convince Peel to allow his name to stand for nomination. Upon the society's return after the summer recess, however, Gilbert's plans for an uncontested nomination hit a snag. Wollaston, to Gilbert's surprise and consternation, put forward the name of Henry Warburton (1784–1858), a close friend and reform-minded member of Parliament. Wollaston may have done this as a tactic to get Gilbert himself to stand for the presidency, and when Peel decided the society was too fickle for his liking and withdrew from the running, that is indeed what happened: Gilbert unwillingly accepted election.

Davies Gilbert, however, was no reformer. One of his first presidential actions was postponing indefinitely the recommendations of the membership committee that would have tipped the Royal Society toward more active membership. In Gilbert's eyes, Herschel and his friends represented the dangerous side of the democratic spirit, a force that could destroy the very institution they were trying to reform. As it turned out, Gilbert's fears were at least partially justified. During his presidency, Babbage and South continued gathering grievances they would eventually use to launch a blistering and public assault on the society, while Herschel looked on from Slough with growing dismay. Initially, though, Gilbert's presidency affirmed Herschel in his decision to retreat to Slough even as his reforming colleagues in London vented their frustrations in increasingly dramatic fashion. When Sabine, who had coordinated with Herschel on measuring the longitude between Paris and Greenwich, was elected to take Herschel's place as Royal Society secretary, South went into a rage so extreme, Herschel told Babbage, that "all other wrath that I have seen in him or other of my friends is mildness in comparison." Gilbert, in an attempt to retain a balance between resisting reform and maintaining the society's scientific credibility, extended what he hoped would be an olive branch with an appointment he was sure would please "every individual member" of the society: Herschel as vice president. Herschel, still resolute in his withdrawal, accepted on terms similar to the ones his father had offered when elected first president of the Astronomical Society: he would be honored, if it was understood that his astronomical work kept him from any official duties. Gilbert agreed, initiating a connection by which Herschel's credibility was linked to society leadership even as Herschel remained personally disengaged.[27]

In the midst of his retreat from the "cabals and contentions" of the metropolis, Herschel retained his positions on the Board of Longitude and the Glass Committee. But he soon received news that deepened his frustrations. On 15 July 1828, Parliament repealed the act that had created the Board of Longitude, disbanding the body that had provided a means of support for research determining longitude for over a century. Responsibility for the *Nautical Almanac* passed to the Admiralty, along with the board's instruments, while its books and records were given to the Royal Society. The board had not been perfect, and, as a member, Herschel had frequently disagreed with how it operated, what kinds of projects it supported, and the quality of the work it produced. But at least it had been an active body within the government pursuing and supporting scientific investigations.[28]

With the dissolution of the board, Herschel's fears regarding lack of governmental support for the sciences in England were confirmed. Dissolving the board immediately jeopardized several major projects. Herschel had recently secured board sponsorship of the reduction and publication of the astronomical catalog produced by the Parramatta Observatory in Australia, the first southern star catalog by British observers and important for improving navigation in the Far East. There was also the forthcoming astronomical supplement to the *Nautical Almanac*, which Herschel and colleagues in the Astronomical Society had been trying to get added for years, the fate of which was now in doubt. Finally, the Glass Committee, a joint project between the board and the Royal Society, now also faced an uncertain future. The end of the Board of Longitude meant all these projects, which Herschel saw as vital to Britain's interests, no longer had funding or government support. Beyond the impact on specific projects, the board's dissolution also communicated the government's perceptions of the value of natural philosophy in general. "What *will* the civilized world," Herschel wondered, "say to the cavalier kind of way in which science and men of science are treated in England!"[29]

When Young, the board secretary, wrote to Herschel asking whether they should continue work on the *Almanac* and the proposed astronomical supplement, Herschel's response was curtly pessimistic. The supplement had been his plan to reform the *Almanac* from within, adding change "cautiously and gradually" and making the *Almanac* a stronger and more useful tool more along the lines of the annual astronomical handbooks produced in Europe. But that plan was pointless now. "It *is* idle for us to attempt competition with our Continental neighbors," Herschel complained to Young, "whether French or German, in matters of science generally. Our day is fast going by. . . . We are rapidly dropping behind the race." His position within the board had been his only official government appointment and the only scientific position for which he was paid. But now he refused the final installment of his salary, telling Young to send it back to the Admiralty. "I should think it foul shame to continue to pocket the wages of a Government which treats its agents so Cavalierly, a moment after the true nature of their relation to it is made apparent." For Herschel, independently wealthy since his father's death, refusal to take his final payment was a protest of how the government treated its scientific experts.

The decision to dissolve the board also changed Herschel's view of his role as an astronomical practitioner. The Parramatta observations, for example, had to undergo long, tedious calculations that would reduce them to useful data. The board had called this "a work of national importance," and as such, Herschel had been willing to do the calculations himself as a "public functionary" discharging a public duty. But the decision to dissolve the board invalidated this perception of the catalog as a public good, and Herschel no longer had any intention of pursuing the project as a volunteer. In terms of astronomy, Herschel now had his own work to do.[30]

The fall of the Board of Longitude confirmed Herschel's fears regarding natural philosophy in Britain. The contrast between what his country was producing and what he perceived abroad was particularly sharp, Herschel believed, in astronomy. Apart from the inferiority of British glass, British observers could not compete with the accuracy and detail of European star catalogs and tables. After Johann Encke released his *Berlin Ephemeris*, for instance, Herschel told a colleague the quality of the work far exceeded anything the British could produce: "While *we* have been shilly-shallying *he* has been doing. England seems fated now to lag in the rear of all her neighbors—at least in science."[31]

The time since Herschel's return from Europe had taught him not to expect the reform of science to come either through government support or the behavior of the members of England's scientific institutions. He had initially responded to this realization by composing "Light," which illustrated his hopes for unity and elegance both in theories of the world and the conduct of natural philosophers—if elegance was interpreted as cooperation and the disinterested pursuit of knowledge. Herschel hoped to eventually rework "Light" into a nonmathematical account for a larger readership, but instead the work that would have been Herschel's popularization of light became a popularization of something else: the scientific life itself. Herschel had recognized the need for science to be seen as a moral pursuit, to transcend the disagreements of its own practitioners. His London discontentment prepared him for a work that would reproach his colleagues who failed to embody the rational morality that science should engender. If Herschel could not reform natural philosophy from within its institutions, he would go outside those institutions, pushing science beyond its previous social and cultural boundaries as his father had pushed astronomy beyond the confines of traditional positional astronomy. He would not reform the Royal Society but the practice of science itself: opening and outlining its methods to a growing middle class and showing that its pursuit did not live exclusively in the gilded halls of Somerset House.

CHAPTER 7

STABILITY OF
THE SYSTEM

The curfew tolls the knell of parting day,
The lowing herd wind slowly o'er the lea,
The ploughman homeward plods his weary way,
And leaves the world to darkness and to me.

—THOMAS GRAY,
"ELEGY WRITTEN IN A COUNTRY CHURCHYARD," 1750

In the middle of the eighteenth century, the English poet Thomas Gray
(1716–1771) completed a poem he had been at work on for a decade
and that would eventually become one of the most popular poems in
the English language. There are various claims for the iconic country
church that provided the setting for Gray's elegy and inspired his mus-
ings on pastoral life and mortality with its overgrown graveyard. One
strong possibility for the "ivy-mantled tow'r" of Gray's poem is Saint
Laurence Church in Slough, across the fields from Eton College where
Gray passed some of his most pleasant boyhood years and where, much
later, John Herschel was briefly enrolled as a pupil.

The solid flint structure of Saint Laurence Church dates back to the
twelfth century and is one of the oldest surviving buildings in the town
of Slough today. The church is a classic example of Norman architecture,
with narrow windows and low slate roof, and remains largely unchanged
since Gray's time, though in the early twenty-first century the church received a bequest to add a modern stained-glass window commemorating
William Herschel's discovery of Uranus. The changeless solidity of Saint
Laurence Church was a fixture in John Herschel's childhood and early

adult years. He was baptized at Saint Laurence soon after his birth, with his father's scientific friend Sir William Watson standing as godfather, and as a child and young man he attended services in the church regularly. The church grounds would ultimately become the resting place of his mother and father, both of whom were buried in a vault at the base of the tower.

Saint Laurence, architecturally and metaphorically, symbolized the stability of the established Anglican faith. In a period that saw the chaotic effects of the French Revolution, an intensification of spiritual life with the evangelical revival and Methodist movement of the late eighteenth century, and increasing political assaults on established state religion, stability for many during Herschel's childhood and early career became a tenant of religious faith. The universe itself strengthened this view by embodying stability in its own structure. The Newtonian view of an orderly planetary system provided the exemplar of elegant stability and a powerful argument for a creating and sustaining God. Newton himself had acknowledged early in the *Principia* that the "most elegant system of the sun, planet, and comets, could not have arisen without the design and the dominion of a great and intelligent being." This idea was developed further through the writings of popular natural theologians, including the philosopher and clergyman William Paley (1743–1805), whose *Natural Theology, or Evidences of the Existence and Attributes of the Deity*, was published in 1802 and as part of the Cambridge curriculum provided a framework for Herschel's generation to understand and appreciate the orderly relationship between God and nature.[1]

The idea of a stable astronomical system created and maintained by God was powerful rhetoric for those wishing to maintain stability in the social and political landscape as well. As revolution convulsed France and sympathetic uprisings threatened Britain, political chaos was seen the result of radical secular reform. It was no accident, it seemed clear, that the secularism and materialism of French natural philosophers had resulted in political turbulence. State-sanctioned Anglican Christianity, the ancient faith of the countryside maintained in hundreds of venerable churches like Saint Laurence in Slough, was the essential stabilizing force in the nation's fabric. It was reassuring, then, that the general consensus among British natural philosophers was that scientific pursuit did not undermine this faith but rather confirmed and supported it.[2]

Ironically, a possible challenge to this alliance between theology and astronomy came from the very stability of the planetary system that astronomy revealed. Laplace's demonstration that gravitational perturbations among the planets were not destabilizing in the long term, far from being seen as confirmation of the created order of the system, was taken by some (including Laplace himself) to indicate the solar system was *naturally* stable, needing no higher power to maintain its order. Though William Herschel may not have been bothered by Laplace's answer to Napoleon's question—posed at their meeting in 1802—of where God was in this system, he was dismayed to learn that his own work was taken to support an atheistic interpretation of Laplace's ideas, especially on the Continent. If the solar system was naturally stable, then there could also be a natural explanation for the evolution and origin of the stars themselves. Though William's theories of how stars formed were independent of Laplace's theory of the origins of the solar system, to many they seemed logically connected, and William's ideas were often taken as confirming Laplace's view of a universe evolving without need for divine intervention. Against this perception, John Herschel continually stressed that his father's work should be taken as confirming rather than undermining traditional religion.

Herschel, however, had not always been a supporter of the established Anglican faith. As a student at Cambridge, he had decried both evangelical fervor on one hand and the abuses of the established church on the other. During this period, government positions were closed to anyone who was not Anglican, and dissenters (as non-Anglicans were called) could not obtain university postings. This would change during Herschel's lifetime, but upon his graduation from Cambridge—even as his father urged him to consider a vocation in the church—Herschel voiced the fervent "wish & prayer, that religion, as *established by law*, may never entirely usurp the superiority & control over religion, established by nature." Religion established by nature was rational and not necessarily based on personal piety or devotion. This view disappointed his friend James Grahame, who as a faithful Scottish Presbyterian often bemoaned Herschel's lack of personal devotion.[3]

Travel in the years after Cambridge softened Herschel's views toward Christianity somewhat. His narrow escape from an overturned carriage along a mountain road in Italy forced him to think seriously about divine providence. On the same trip, his aversion to what he perceived as irrational religion was affirmed by the fervor of religious festivities in Palermo and superstitious traditions he found disgusting and demeaning. The rational belief owed to God, Herschel felt certain, had nothing to do with crawling on one's knees across the floor of a cathedral, licking the tiles in penance. On the other hand, attending a papal blessing in Saint Peter's Square he noted both its emotional effect as well as the way the immense crowds, which numbered several thousand, departed in peaceful order upon its conclusion. Institutionalized Christianity, even Catholicism, could have a rational and ordering effect on society.[4]

Herschel was especially drawn to the stabilizing aspects of religion in the home, anchoring and ordering the life of the family. In his rambles alone through the Austrian Alps, he had been greatly affected coming into villages in the evening as all business ceased and he heard prayers read and hymns sung in each house he passed. "I have more than once longed for your presence," he wrote to Grahame, whose pious feelings had always been stronger than his own, "when I have heard the same sounds of prayer and praise ascending from lonely cottages in thick woods or among savage rocks." Christianity could be seen as the bulwark of a quiet life, from family prayers to parish worship to society at large, and Herschel's religious practice eventually bore this out. After his father's death, if he was in Slough on Sundays, he was either attending liturgy at Saint Laurence or reading prayers aloud with his mother at home. Ultimately, his reform of natural philosophy would emphasize and enshrine support of this rational, stabilizing view of religion as one of the greatest goods provided by the pursuit of science.[5]

If the stability of systems in the sidereal universe supported a rational, stabilizing Christianity, those systems should be mirrored by processes on Earth's surface. Even as Herschel worked to complete his father's survey of nebulae and created additional catalogs of double stars discovered along the way, he continued to pursue geology, traveling to southern France to follow up on his observations of the active volcanoes of Italy with a study of the region's famous ancient ones. In particular, Herschel wanted to see the evidence of this region with his own eyes to decide between a view of Earth's past shaped by cataclysmic events (such as enormous floods or volcanic conflagrations) or a history better understood by slow and steady accumulation of gradual changes still at work on Earth's surface. In other words, Herschel hoped to determine whether the stabilizing paradigm of the planetary system extended to the surface of Earth itself.

AUVERGNE

In August of 1826, traveling when his Royal Society responsibilities in London ebbed during the summer season, Herschel set off with his servant James on his final European scientific pilgrimage. After an obligatory stop in Paris, where he dined at Cuvier's home in the company of Charles and Georgiana Babbage, who were also visiting Paris, he pushed on to Auvergne, a rural district containing a range of low mountains called Chaîne des Puys. Auvergne had long been a target of French geologic studies, as its landscape of extinct volcanoes offered a means of addressing both questions regarding the formation of certain rock types as well as the relative ages of lava flows and river valleys. The end of the long Napoleonic wars opened the region to British geologists as well. Buckland had followed Daubeny, whom Herschel had met in Sicily, with a tour of the region in 1820. French geologists had used the region for evidence of slow processes, but Buckland felt the landscape was better interpreted as evidence of a deluge. The year of Herschel's trip, Daubeny had published his *Description of Active and Extinct Volcanoes*, which followed *Considerations on Volcanoes*, published by the British geologist Scrope the year before. Whereas Scrope argued for a gradualist, uniformitarian explanation for volcanoes, his book was viewed by other British geologists as too speculative. Daubeny's work focused more on description and less on theory and offered an explanation more in keeping with Buckland's views. Herschel would wade into this discussion by seeing the region for himself.[6]

Based on his ascents of Etna and Vesuvius, Herschel had no doubt the slopes in this region were indeed volcanoes extinct for centuries, their rounded domes in fact craters that had erupted before recorded history. Though such a landscape might have been punctuated by violent eruptions in the past, Herschel became convinced that overall, the contours of the region he rambled through with James were shaped by the same volcanic effects he had witnessed firsthand at Etna and Vesuvius operating slowly over millennia. Herschel's view of Auvergne was of evidence for continuity and stability, not cataclysmic change. His observations, he told a friend, made him an "ultra-Huttonian." The landscape, with its mix of volcanic cones and river valleys carved through them, was shaped, Herschel relayed to Babbage, "certainly not by *distortion* (i.e. violent & sudden causes) but by the *slow* action of causes now at operation."[7]

This conception of geological stability and gradual change conformed with Herschel's continuing astronomical work. As he spent more time observing nebulae and double stars, he became convinced that in the sidereal heavens as well as in the solar system and on Earth's surface, stability reigned. If there was an evolutionary progression in the heavens, like his father believed, it was not one of cataclysm and chaos but gradual transition. The possible coalescing of nebulae into stellar systems was slow enough to require the most precise and careful observations to perceive. The velocity of double stars in their orbits around their common center of gravity might be immense, but those twin stars were bound in stable configurations. Herschel would eventually link this view of stability of the heavens directly with the system of Earth in a paper he presented to the Geological Society in which he argued that gradual changes in Earth's climate might be an effect of slow changes to the shape of Earth's orbit. His view of heavenly stability would be challenged by observations Herschel made of variable stars at the

Cape of Good Hope, but before that, stability seemed the rule of the cosmos—as well as of the social and spiritual affairs of life. A rational faith that cultivated and enshrined stability, while avoiding the pitfalls of emotional enthusiasm on one hand and the abuses of state-sponsored institutionalized religion on the other, became Herschel's formula for his understanding of Christianity.[8]

Herschel also experienced this sense of stability and tranquility woven into the domestic life of some of his closest friends. The stabilizing aspects of a rich family life, for example, were essential to his longtime colleague Babbage. By the time Herschel returned from touring the ancient volcanoes of Auvergne, Babbage and his wife, Georgiana, had five children, the first of whom they had named Benjamin Herschel in honor of their friend. Babbage's relationship with his father, who had initially opposed his marriage, had been patched up, and Herschel enjoyed the warmth of the Babbage home and his close friend's obvious domestic happiness on his frequent visits. Despite his outbursts against London's scientific institutions, Herschel knew he could count on the stability of Babbage's home and family as a counterweight to his vitriol. When this happiness and stability was shattered by sudden tragedy, however, not even Herschel could prevent the worst of a grieving Babbage's outbursts.

ENGINE OF CHANGE

In the years since Cambridge, while Herschel's scientific credibility continued to grow, Babbage's path had been more difficult. Though initially supported by his banker father, this support had been temporarily lost when Babbage married, making Babbage eager for a professional position and financial security. While Herschel traveled Europe on his grand tour, Babbage had taken the position of accountant for a new life insurance company, putting his mathematical mind to the task of calculating insurance premiums and investments, only to witness the company fail through mismanagement by the other partners. Venting his frustrations with a geological tour, a capsized boat cost him all his supplies and the notes he had made for a book he was planning to write on insurance calculations. When Herschel learned of Babbage's latest run of ill luck, he commiserated that his friend's plans often had "a nasty vicious way of flying out from under him." Babbage, on the other hand, took this in stride and told Herschel he was planning on purchasing a lottery ticket, as it was now statistically certain his luck must change.[9]

There was one project in particular Babbage was confident would establish his scientific career: his calculating machine, or difference engine. Much as Herschel and Babbage had worked in their early papers to develop methods for simplifying and standardizing complex equations, in 1822 Babbage had begun work on a calculating machine that would mechanize these mathematical tasks. The idea came to Babbage after he and Herschel had spent a day working on the tedious equations needed to reduce astronomical observations. Herschel had wondered idly why steam, which was transforming the world of transportation and manufacturing, couldn't be harnessed for calculations as well.

Babbage soon devoted himself to designing and constructing a mechanical device that could do exactly what Herschel had suggested. Because of the funds required for precise machining of the parts involved and his certainty that such a device would save hundreds of hours of human calculation time and expense, Babbage believed the British government should financially support his work. His difference engine would, for instance, increase the accuracy and speed of calculating data for the *Nautical Almanac*, essential for commerce and navigation. Due in large part to the support of Davies Gilbert, who despite the frustrations of the reformers in the Royal Society still functioned as the primary connection between the society and Parliament, this support materialized in 1823 with an unprecedentedly large government grant of almost two thousand pounds.

The difference engine quickly caught the imagination of the wider scientific community and became Babbage's pathway to acclaim and eventual notoriety. When Herschel traveled throughout Europe, the astronomers he met with wanted to hear the details of Babbage's device and were impressed by what such a machine implied for their work. The difference engine was another step along the road of reforming mathematics that Herschel and Babbage had begun in Cambridge. Their analytical revolution provided methods to develop complex new equations, and Babbage's machine would provide a means of solving them quickly and efficiently. With the difference engine, analytic revolution met industrial revolution.[10]

Construction of the engine and expansion of his original plans for it began consuming more and more of Babbage's time and efforts as pressure mounted to make good on his promises and fulfill the obligation placed on him by the government's funding. At the same time, Babbage grew frustrated by the situation with Humphry Davy and the Royal Society. Losing the chance at the Royal Society secretary position was a disappointing setback and personal embarrassment, as was his failure to secure a Royal Society medal. Despite the government's initial financial support, Babbage became convinced that interests in the Royal Society and Parliament were actively working against him or were too ignorant to recognize the true importance of his work.

As tensions mounted, Babbage's domestic situation helped reign in his more vindicative responses and, along with Herschel's occasional pleadings for calm and patience, tempered his impulsive and angry reactions. Babbage's marriage and home life anchored him and provided stability and support even though he felt persecuted or unappreciated outside of the home. Those moorings, however, suddenly gave way when, in a span of only months, Herschel's oldest and closest friend lost his father, his ten-year-old son, and finally Georgiana, his wife of thirteen years, and their newborn infant. Babbage was devastated. His domestic universe was destroyed, and he never truly recovered from the tragedy. Work on the difference engine, which was already running into problems due to Babbage's demands on the artisans producing the parts, ground to a halt. Herschel, anxious to help his friend, dropped his own projects and, immediately after Georgiana's funeral, took Babbage to Ireland for a geologizing trip. As they clambered over the Giant's Causeway and conducted experiments timing echoes under the recently constructed Menai Suspension Bridge, Herschel contended with thoughts of mortality and his friend's anguish.

Upon their return, Babbage immediately departed England, following Herschel's footsteps for a long solo trip through France and into Italy, leaving his surviving children in the care of relatives and putting the continuing construction of the difference engine under Herschel's supervision. His long absence put Herschel in the awkward situation of carrying out Babbage's designs in the face of diminishing funds, confused instructions, and slow correspondence—as well as being called on to defend Babbage's work when a letter was published in a London newspaper asking what had become of the government funds Babbage was originally awarded. With Babbage away in Naples, Herschel replied with a public letter in the *Times* defending both Babbage's project and his conduct.[11]

Babbage would remain abroad until the end of the following year. While he was away, the controversies he had stoked with the Royal Society in large part died down. Yet he returned from the Continent with the same frustrations regarding lack of support that had been building for years, reinforced by the same contrast Herschel had witnessed in the support of science in France and the German states, and he would confront these mounting grievances without Georgiana's equalizing influence. Instead, Herschel hoped to be able to provide his returning friend a different kind of stability. In Babbage's absence, positions had opened up for professorships at Cambridge. If Herschel could exert his influence, it might be possible for Babbage to step into one of these university positions. Perhaps, if Babbage was professionally and vocationally a part of the establishment, he would do less to rage against it and, like Herschel, begin to see the merit in reforming it from within. There would also be delightful irony in the Cambridge graduate who had bucked convention with his undergraduate act of disputation and worked with Herschel to bring about mathematical reform returning to the university to take one of its most prestigious chairs. The only problem was that first Herschel would have to convince Cambridge that despite his own previous efforts for one, his scientific path must now remain independent of any formal university position.

AN INDEPENDENT AMATEUR

Though Herschel had long before lost interest in a position as a Cambridge professor, he remained formally connected to the university through his fellowship at St John's College, despite the fact that fellowships usually entailed remaining an unmarried resident of the college and eventually becoming a clergyman. Herschel was still unmarried, but that the residency requirement was never enforced was a sign of his status and growing influence. When it became clear he did not intend to fulfill the vocational requirement either, he was offered a law fellowship instead, meant to support those studying for the court instead of the church. He declined, stating that if he ever chose to resume the study of law, he wanted the fellowship to remain available for those who had financial need. Nonetheless, the college voted him another fellowship, this time one with no responsibilities at all except to be willing to return in five or six years if requested. Herschel agreed, saying he would visit Cambridge when possible but pledging to resign the fellowship "whenever a meritorious individual to whom it may be of consequence" might need it. The fellowship ensured that Herschel remained formally connected with the governing of St John's even though he remained in London or Slough. Herschel also collaborated with Cambridge fellow and mathematician George Peacock in a movement to establish a new observatory at the university, staying up-to-date on its progress and providing feedback on plans.[12]

As professorship opportunities came open, Herschel was forced to articulate his own approach to the pursuit of science in explaining why he would not seek them. When the Lucasian Chair had opened in 1826, his colleagues urged Herschel to put his name forward, confident it would be his if he wanted it. "I shall do more for science as an independent amateur than as a Professor of any particular branch or department of it," he told one correspondent, declining to submit his name for consideration.[13] To his Cambridge friend Whewell he was more explicit, assuring Whewell that his decision did not come from resentment from failed previous attempts. Rather, he no longer wished to devote himself "exclusively or par excellence to any one branch of Science." He now preferred "physical to mathematical science," a preference that was "increasing as I get to know more of one and less of the other." Herschel had originally left Cambridge with a passion for pure mathematics, but the years since had transformed him into a natural philosopher who saw mathematical analysis as a tool for investigating everything else—the heavens, mountains, crystals and lenses on his laboratory bench, the mysteries of light. A chair in mathematics was not broad enough to contain Herschel's view of natural philosophy.

Herschel also believed it was better for him to pursue his work free from financial incentive or institutional obligation. He wanted his endeavors, he told Whewell, to be considered the work of a devoted amateur rather than "a matter of duty and profession." For Herschel's developing ideas on the moral virtue of science, science pursued because it was one's job was no better than a person being honest or patient because they felt it was part of their professional duty. That did not mean professors could not be natural philosophers, but Herschel himself could afford the freedom to ensure (and ensure that others knew) that his scientific virtue was unencumbered and absolutely free of duty. His contributions to the study of nature would not be "giant inroads into great branches of human knowledge—but rather to loiter on the shores of the ocean of science and pick up such shells and pebbles as take my fancy for the pleasure of arranging them and seeing them look pretty." This sense of free play, which Herschel felt was essential to his scientific practice, could only flourish if not bound to any professional or institutional position.[14]

Herschel continued to articulate this stance when he was urged to take the Plumian professorship, one of Cambridge's two astronomy chairs. It was a chance, Herschel admitted to a friend, to leave the squabbling of London behind and be free of "the annoyance and 'badgering' of an impertinent and imperious public which renders *national* responsibility so very irksome," likely referring to his frustrations with the Board of Longitude. Ultimately, though, Herschel rejected the offer for the reasons previously given: "I feel that I shall work more to purpose when not compelled to work by a sense of public responsibility, and that the little I may contribute to the progress of Science will shew to more advantage (at least will be *more good-naturedly appreciated*) if done from mere bent of natural inclination, than as official duty." It did not matter that this particular chair was in astronomy. "I assure you," he wrote in declining the invitation to apply, "I find stargazing hard duty enough as it is without adding to it the sense of compulsion, without bending myself not to quit for so many months in the year, and to lecture on its first principles for so many hours in the week." Even observing nebulae and double stars would lose its appeal if done from obligation.

An additional passage in a draft of this letter (struck out by Herschel but still legible) provides additional insight into his ideas of how his scientific pursuits would be perceived: "My wish is that what little I may contribute to the advancement of science may go down to posterity as the contribution of one who at least was disinterested in his exertions. . . . After my death, I would give myself the advantage of having laboured from a sense of the beauty and utility of science and not for the collateral advantage of forwarding my interests."[15] Thanks to his financial independence, Herschel had the freedom to pass up opportunities he had coveted when he was younger. But he couched his decision in moral terms. Science—and by now Herschel was using this term in its modern sense, to describe the physical sciences including astronomy, chemistry, and physics—was most pure, most credible, and most worthy of praise when a disinterested pursuit. His scientific legacy, he believed, would be threatened if it was perceived as being pursued either from a sense of obligation or remunerative gain.

Because Herschel refused to put himself forward for the Plumian professorship, the position went instead to the astronomer George Airy, and because Airy had held the Lucasian Chair, this position now opened yet again. Herschel was a third time approached, and he used the opportunity to suggest Babbage. Thanks in large part to Herschel's advocacy, Babbage gained the position, finally giving him the professional standing he had sought for years. Yet to Herschel's dismay, after securing the position Babbage lingered for several additional months in Europe before returning to Cambridge. Moreover, when he finally did take up his university post, Babbage never took its responsibilities seriously, though he held the post for the next eleven years. What Herschel hoped would bring Babbage a measure of stability after his personal losses instead simply gave his friend an additional platform to resume attacks on the Royal Society, the government, and the situation of science at large in Britain.

Herschel, on the other hand, remained steadfast in his withdrawal from disputes involving the scientific community. His consistent denials of positions at Cambridge were in keeping with his determination to pursue his work in the context of his own domestic sphere. The controversies within the Royal Society had convinced Herschel that the home, not the lecture hall or meeting room, was where science could be pursued most effectively, and his decisions against Cambridge confirmed this choice. Yet Herschel's ideas of what kind of home this would be were soon transformed when he met the daughter of a Presbyterian minister and was forced to reevaluate his perceptions of both domesticity and evangelical piety. The scientific ideal Herschel had been developing since Cambridge gained an important additional layer through the partnership he soon formed, and the home he would build, with Margaret Brodie Stewart.

MEETING THE STEWARTS

When Herschel returned from Ireland with Babbage, he found his old friend James Grahame had temporarily moved to London, where Grahame's daughter, Matilda, was attending boarding school. Herschel had just relinquished his position as Royal Society secretary and was changing lodgings in London. As Grahame was looking for a place to live as well, Grahame suggested they room together and create a household as confirmed old bachelors (or, in Grahame's case, as a widower). He knew, he told his friend, that as "shepherd of the stars" Herschel had to balance his observing time at Slough with a presence in London, but Grahame thought the domestic arrangement would be ideal. Herschel did not take Grahame up on the offer, but Grahame's presence in the city soon played an essential role in establishing Herschel's own domestic stability.[16]

Grahame had been encouraging Herschel to get married for years and eventually came to see marrying Herschel off as something of a personal project. He kept Herschel appraised on what he heard about him in popular society, noting that as Herschel's fame grew, so did his reputation as an eligible bachelor. At the same time, Grahame was worried Herschel would acquire solitary habits in his long bachelorhood that would make marrying more difficult down the road. Herschel's aunt Caroline urged him along the same lines, telling Herschel to marry before he grew too "old and cross." Likewise Grahame needled him that "there are many good girls in the world. But how are you to find them out?" The implication was that Herschel needed to mix more in society and less among extinct volcanoes. But Herschel's time was too occupied to pursue any sort of courtship. His evenings were spent continuing his father's survey of nebulae and his days were taken up with writing. Grahame realized it was up to him to introduce Herschel into someone's social circle and hope for the best.[17]

Babbage and Grahame had already outlived their spouses, but permanent bachelorhood was always an option, and Herschel began to resolve himself to the idea. Sedgwick, for instance, remained unmarried in Cambridge, free to give lectures and pursue his geological studies without the need to provide for a wife or family and apparently perfectly happy. Whewell was a bachelor fellow as well and would not marry until 1841. On the other hand, should Herschel decide to marry, the financial arrangements that had plagued and ultimately derailed his first engagement were no longer a problem. Money was not an issue. When Grahame ran into financial difficulties, Herschel urged him to take advantage of Herschel's wealth. "I am rich," Herschel wrote his friend, and "whatever notions you may have formed to the contrary . . . I am unmarried and *likely to remain so*."[18]

None of Grahame's earlier plans for Herschel's marital happiness, including a scheme to have Herschel marry his niece, had come to fruition when Grahame moved to London and they resumed a friendship that had for many years been maintained largely through correspondence. As he settled into the London social scene, Grahame connected with a network of Scottish immigrants associated with St. Marylebone Parish Church on the city's west end. In particular, Grahame resumed his acquaintance with Emilia Stewart (1780–1855), the widow of a Presbyterian minister named Alexander Stewart (1764–1821) who had been a close friend and mentor of Grahame's in Edinburgh. Emilia lived in Nottingham Place, just around the corner from Marylebone church, with her children, who included two young daughters.[19]

The Stewarts were active members of the Marylebone church and friends of Charles Wesley Jr. (1757–1834), who lived nearby with his sister Sarah Wesley (1759–1828) and was employed as the church organist. Charles was son of Charles Wesley (1707–1788) and nephew to John Wesley (1703–1791), both famous as founders of the Methodist movement. In late February of 1827, Grahame told Herschel he was going to visit the Stewarts, who were taking him to meet Charles Jr. and his sister. Though Wesley was an orthodox Anglican who appears to have had little of his father's religious fervor, an earnest evangelical like Grahame was happy to meet someone so close to the center of the Methodist movement. Before long, Herschel would become caught up in this evangelical social circle as Grahame used his friendship with the Wesleys and Stewarts as another opportunity to match Herschel up with a potential wife—and perhaps save his soul.

Even as Grahame was building connections in London, Herschel was making good on his retreat to Slough and devoting himself more exclusively to astronomical observations. Grahame visited him there just as carpenters began constructing a room to house the equatorial telescope Herschel had purchased from South, but throughout the spring of 1827 Herschel consistently declined invitations to London, including social invitations from Grahame. That summer, as Herschel was left fuming by the Board of Longitude's sudden dissolution, Grahame left London for Nantes, France, where his son, Robert, was attending school. He begged

FIGURE 7.1: Sir John Frederick William Herschel, *Netley Abbey. Southampton. East Window & South Transept*. Graphite drawing made with the aid of a camera lucida, unframed: 7 3/4 × 12 1/2 in. The J. Paul Getty Museum, Los Angeles, Gift of the Graham and Susan Nash Collection.

Herschel to check frequently on Matilda, who remained in boarding school in London. He also asked Herschel to look in on his good friends the Stewarts, and in September, Herschel's diary records that he dutifully followed Grahame's wishes and called on Emilia and her family in their apartment in Nottingham Place.

Though Herschel had met the Stewarts at least once prior in Grahame's company, the September visit was a turning point. The following week, when Grahame returned to London for a short stay, the two friends rowed by moonlight from the town of Southampton to the ruins of Netley Abbey and passed an hour in "whispered confidence in the main aisle under the ash trees with the stars looking in above." The crumbling remains of the monastery, in ruins for the past century, formed an atmospheric backdrop for a long conversation on Herschel's romantic prospects. He was, he admitted to Grahame, quite taken with Margaret, the eldest of the Stewart daughters. Yet he was also hesitant. The disaster of his previous engagement haunted him, and he had not considered marriage seriously since that abortive attempt more than seven years before. There was also a significant difference in their ages, not enough to be remarkable for the time but enough to give him pause.

Margaret, or Maggie as she was usually called, was only nineteen; Herschel was thirty-six.[20]

Grahame returned to Nantes, where he coached Herschel on his potential courtship via correspondence. Grahame also fretted about the fate of Charles Wesley Jr., whose sister died that year and who, Grahame feared, was living with relatives taking advantage of the distracted musician's financial naivety. The connection with Wesley proved a complicating factor in Herschel's interactions with the Stewarts, who like Grahame felt obligated to help care for the aging musician. The eccentric Wesley had become part of Herschel's social circle as well; he and his sister had visited Herschel and his mother at Slough just prior to Sarah Wesley's death. As Herschel considered his next steps regarding Margaret, the son of England's most famous astronomer frequently crossed paths with the nephew of England's most famous religious reformer.

Herschel looked for ways to get closer to the tightly knit Stewart family through another family tradition: music. His father, after all, had been a musician before becoming an astronomer, and his aunt, before giving it up to assist her brother in his astronomical endeavors, had been a vocal soloist on the way to a promising professional career. Herschel himself was a passable flutist, and the Stewarts had recently purchased a brand new piano that needed accompanying. He seized on this opportunity as an excuse to visit and accompanied Margaret as she played (though he likely did not share with the minister's widow and her family about the time he had been disciplined at Cambridge for missing church to spend his Sunday morning playing the flute).

Herschel, who was effectively an only child, came to appreciate the warmth of the Stewart household almost as much as he did Margaret's company. His only sibling, his half brother Paul, was grown and living away from home when Herschel was born, and Paul had died soon after. As the only child of two parents who were relatively old for child-rearing, his family life had been in some ways a lonely one. "I stand almost alone in the world," Herschel wrote to Emilia, explaining the warmth he felt in the company of her family. "Hardly any man has fewer family ties. Except my mother, two first cousins . . . are my only near English relative[s]." The Stewart residence in London was a change, full of a domestic warmth he had only experienced in the homes of others.[21]

Of course, this was all viewed through the lens of his growing feelings for Margaret, "a most delightful creature," he told Grahame, whose "rational manner and sincere unaffected attention and kindness" had quickly won over Herschel's elderly mother as well. The only problem, perhaps reminding Herschel uncomfortably of their age gap, was that she would not leave off calling him "sir," an "odious" monosyllable that may have also hinted at the social gap between him as a wealthy gentleman philosopher and her as the daughter of a minister's widow. He continued to call on the Stewarts regularly throughout the autumn of 1829 and began attending services with them at Marylebone church on Sundays, where they heard Wesley play the organ. "Such are the hours," the soon-infatuated astronomer recorded in his diary, "we live for!" It also quickly became clear to Herschel that the Stewarts' pious faith and religious practice would force him to become more serious about his own.[22]

It is unclear what Margaret thought of the sudden introduction of this reserved, older gentleman bachelor into her life. Herschel was well known in scientific circles, and there was of course the fame of his family name, but it was not clear he had the personal qualities Emilia desired in a husband for her daughter or that Margaret wanted in a husband for herself. Herschel represented financial security, and his careful courtesy impressed the family, but for an evangelical family like the Stewarts, the household, even more than the church, was the center of Christian nurturing. Parents were the primary Christian teachers of their children, and the home was the setting for devotion and prayer. As an evangelical woman of this period, maintaining a Christian home and educating Christian children (as well as supporting a Christian husband) would have been central to Emilia's identity and her hopes for Margaret. It was not clear whether Herschel was the kind of person with whom this home could be created.[23]

There also remained the issue of Margaret's age. Herschel recognized he was considering proposal to someone who had never been courted before and whose social experiences were significantly different from his own. He decided to err on the side of propriety and caution, approaching Margaret's mother first and discussing his intentions without Margaret's knowledge. Grahame thought this method best as well, writing to advise Herschel to formally approach Emilia on the very day Herschel had made up his own mind to do the same. Grahame also urged Herschel to pay closer attention to how he dressed, noting that though Herschel was handsome, it never hurt to make a greater effort. When he took the time to dress carefully, Grahame spelled out to his distracted friend, "how much your appearance is improved!"[24]

Herschel called on Emilia on Sunday, 16 November, after attending service at Marylebone church on his own. Emilia was ill and could not meet with him, so he returned to Slough disappointed. He came back to London and visited the Stewart apartment a few days later, where the conversation went as he hoped. Emilia protested that her family had not a penny to their name, and Herschel assured her the fact was irrelevant to his intentions. Emilia ultimately expressed her support but wanted to make sure Herschel's mother knew and approved. She agreed to say nothing to Margaret but allowed Herschel to write her daughter a letter stating his feelings, which she would pass along to Margaret. Emilia would also write to Margaret's legal guardian, who had stewardship of Margaret's financial affairs after the death of her father, as details of the arrangement needed to be made with him.

Herschel was elated. He left the conversation with high spirits and "pleasant daydreams" but walked from it directly into the middle of an awkward situation with the absentminded Wesley. Though Wesley was a musical prodigy who could play any piece of Handel's from memory and had often performed for the king and his family, the musician had lived his adult life with his grown sister, who had set his outfits out for him each morning as he was otherwise liable to forget articles of clothing. Brother and sister had subsisted together on his small salary as organist, but Sarah's death raised the question of where Wesley would now live and who would care for him. Herschel had offered for him to stay at Slough and arrange his father's music, but it had ultimately been decided that Wesley would board with a Mrs. Mortimer until the reading of his sister's will, which hopefully included financial entailments for Wesley's support.[25]

Emilia was concerned about these arrangements, so Herschel left his meeting with her having promised to call on Wesley and arrange a time to accompany him to the reading of the will. Arriving at Mrs. Mortimer's, however, Herschel found, instead of the "poor old man," news that Wesley had not a half hour previously departed in "company with three female acquaintance whose names I could not learn who carried him off in a hackney coach *with all belonging to him*." The women had ransacked Wesley's belongings and cleared his apartment "of all his worldly goods but his old, white hat, and leaving on the table a short scrawl in his handwriting addressed to Mrs. Mortimer but hardly intelligible." It looked as though the musician had been abducted by relatives hoping to gain access to whatever he was promised in his sister's will. Herschel regretted not locking Wesley up until the will was disposed of, telling Grahame the inheritance could have bought an annuity that would have provided for Wesley the rest of his life. As it was, the location of "poor old Mr. Wesley" was then unknown. Grahame, still in France, could only pray that "the God who his father & uncle served so zealously" would "protect the weak old man from his own folly."[26]

Apart from interruptions caused by Wesley's domestic drama (which perhaps served as a warning to Herschel of the dangers of prolonged bachelorhood) a more serious concern to his hopes was the sudden illness of Isabella, Margaret's younger sister. After his conversation with Emilia regarding his intentions, Herschel hoped to call on the Stewarts as frequently as possible. Yet after only a week he was asked to reduce his visits on account of Isabella's health. He obliged but utilized his scientific connections to send something he hoped would help her feel better. Giovanni Plana, one of the Italian astronomers Herschel had stayed with on his grand tour, had recently sent Herschel his published survey of the peaks of the Piedmont, which included drawings of the mountains surveyed. The drawings, Herschel told the Stewarts, offered the only accurate view of Alpine scenery available in England, and he sent the work to Isabella with instructions on how to best view the pictures to reproduce the effect of being in the mountains. Herschel had shared music and conversation with the Stewarts; here was an opportunity to share some of his international credibility as well.[27]

Concern in the Stewart household over Isabella's health was genuine, but there was also lingering concern about Herschel's piety. It was one thing to be the leading natural philosopher in Britain, well traveled, wealthy, and generous, but a sense of religion as a moral duty, a rational obligation, and a stabilizing social force was not enough for the matriarch of this evangelical family or, it seems, for Margaret herself. The Herschel who as a young man had eschewed both organized religion and the emotional fervor of evangelicalism now had to make his reserved, rational, and private faith appear in the best light possible. The Stewarts would not be satisfied by vague pieties; they hoped for a husband for Margaret with genuine Christian faith, and in particular a personal knowledge of Christ as savior.

Once again, Grahame stepped in to assist. In a letter to the Stewarts, Grahame testified to Herschel's spiritual character. He had known Herschel, he told the family, since they had met in Scotland in 1810 and had lived for a time with him in college. Herschel, he assured them, was "a man of pure untainted morals, of the highest integrity, generosity & honour." As far as his specific theological beliefs, Grahame hedged. "I believe him a Christian," Grahame maintained, "though I have often wished that his views of Christian doctrine were more distinct." There was room for optimism, though, as Herschel's religious views seemed "lately to be acquiring more distinctness," no doubt through the influence of Emilia and Margaret and the services he was regularly attending with them. In any case, Grahame was certain what was needed to ensure Herschel's salvation: the only thing the natural philosopher lacked was an "elegant and Christian wife."[28]

In the weeks following his initial conversation with Emilia, Herschel waited patiently through Isabella's illness before finally composing a letter to Margaret in which he confessed his feelings and his hopes that she might feel the same. In the enclosing letter to her mother, Herschel expressed what drew him to the Stewarts: "Were I to choose a family I would look to your fireside, where peace and purity, and innocence seem to have fixed their home. Such a circle I never entered. Surely God is among you. I say not this with reference to any peculiar religious tenets—but where I see such motives producing such results I cannot refuse to acknowledge their origin nor help wishing that in future your paths might be my paths and your ways my ways."[29] Herschel was not ready to go as far as embracing "any particular religious tenets," namely the Stewarts' genuine evangelical piety, but he recognized there was something different about the warmth he found in their home. He assured Emilia he wanted to create a similar household with Margaret.

Margaret responded to Herschel's letter with courtesy and caution. She told him his letter took her by surprise and she wanted time to consider his proposal. For Herschel this was enough, and he waited with optimism. Two days later, after accompanying Emilia to church, he went with her to arrange an investment of twelve thousand pounds as the basis of a marriage settlement. Herschel wanted this taken care of as soon as possible, recalling his errors with his previous engagement. Arrangements proceeded without complications, and by 11 December, less than a month after he had formally declared his suit, he recorded in his diary that he and Margaret were officially engaged.[30]

The day before Christmas, back in Slough with his mother for the holiday, Herschel wrote Margaret their earliest surviving love letter, the first in a correspondence that would continue whenever they were apart for the next forty years. "Every feeling of my heart seems enlarged & ennobled," Herschel wrote, telling her he was searching through his correspondence to find letters from Grahame in Edinburgh from when Margaret's father was alive and recalling one in which Grahame described her father's saintly character and patience under suffering. The two of them, Herschel and Margaret, both shared religious legacies of admired fathers now departed: Margaret the faith tradition her father had been a minister within, and Herschel his own father's rational faith. From personal virtue to the ordering and educating of a family, to the development of stellar systems and the organization of the universe itself, religious faith for Herschel was part of a harmonious whole. Yet he knew he would need to sharpen his expression of this faith to a finer point. He did his best, closing his first letter to Margaret with a distinctly evangelical turn of phrase: "Acceptance before God through Christ be with you and yours."[31]

If stability was a theme in Herschel's religious belief, his partnership with Margaret would soon become another way of expressing and experiencing this stability. As he had witnessed in his father's work, science was domestic. Herschel and Margaret, like William and his sister Caroline, would create space augmenting and supporting the practice of Herschel's natural philosophy in their home (though those roles and spaces would remain gendered and carefully partitioned). Herschel and a generation of natural philosophers to follow would not leave home to do their science or split time between family and laboratory. Herschel's scientific career was conducted within the home, or at a telescope on the lawn, and once this household with Margaret was established, he would continue to resist responsibilities that removed him from that domestic context. When he traveled to South Africa to complete his father's celestial surveys, Margaret and their young children would come as well, creating an outpost of stabilizing domesticity at the frontier of the British empire. Herschel's practice of science was pursued in the context of a family, and his scientific virtues were domestic virtues, girded by a rational Christian faith married to Margaret's evangelical piety.

As Herschel prepared to enter this new stage of life and begin a family, conflict once again broke out in the Royal Society. Babbage, still reeling from his own personal losses, launched his most public attack yet on the British foundations of science. Herschel would be forced to balance a threat to his continuing astronomical projects and newfound domestic tranquility with the possible triumph of his long campaign for scientific reform: by becoming president of the Royal Society—but he would risk censure from the British public by running against the king's own brother to do so. In the midst of this, he would be given the chance to set out his views on natural philosophy in a book that would turn all of his reforming ambitions stretching back to his student days at Cambridge into an appeal for a new kind of scientific life.

CHAPTER 8

REFORM DEFEATED

I feel that I have no choice, or I should have chosen otherwise than
I have done.

—JOHN HERSCHEL TO CHARLES BABBAGE, 26 NOVEMBER 1830

John Herschel and Margaret Brodie Stewart were married in St. Marylebone Parish Church on Tuesday, 3 March 1829. The wedding took place
in the morning, with Herschel's cousin and his Cambridge friend William Whewell standing as groomsmen. Only a few other cousins and
Margaret's older brothers were in attendance. Her younger sister, Isabella, whom her mother was afraid the early morning air would give
a cold, was kept home. Margaret was given away by her guardian, Mr.
McIntosh. An elegant breakfast for the small party followed, and afterward the new bride and groom presented themselves to Emilia Stewart,
who gave her blessing to her daughter and new son-in-law before they
departed for their honeymoon.[1]

Herschel had not traveled for pleasure in his own country since touring with his parents before leaving for Cambridge. The honeymoon was
his opportunity. He and Margaret had decided they would start their
travels in England, visiting fashionable sites. The newlyweds traveled to
Leamington, a rapidly growing spa town near Birmingham, where they
enjoyed the elegance and quiet. The journey was pleasant, except for a
burst of rain that sent them sheltering in the picturesque Gothic Temple
at Stowe. With Leamington as their base, Herschel and Margaret made
excursions to nearby towns like Cheltenham in the Cotswolds. Here,
Herschel enjoyed views of the Severn Vale and Malvern Hills, which

FIGURE 8.1: Sir John Frederick William Herschel, *View from the Hotel Window, Leamington, Warwick, March, 1829*, camera lucida drawing on paper. Gift of Susan and Graham Nash, courtesy of the Museum of Photographic Arts, San Diego, 2002.040.039.

he felt among "the finest and richest in England," before stopping at a nearby quarry to search for fossils. He was delighted to discover that Margaret enjoyed travel as much as he did. "What now can I tell you," he wrote James Grahame from their trip, "but that your best hopes for me are realized, and that I feel myself more blessed than I can tell how to put into words."[2]

Besides hunting fossils, Herschel and Margaret toured the factories of Birmingham, though to Herschel's disappointment they were unable to view the huge Soho Manufactory constructed by Matthew Boulton (1728–1809), which had pioneered steam-powered mass production in England. They did see glassworks, chemical manufactures, and a porcelain factory, however, all industrial marvels considered as important to tourists as the natural and historical sights. After a return to Cheltenham and a visit to Gloucester Cathedral, Herschel and Margaret traveled along the River Wye, a popular route with additional romantic landscapes to take in. At Monmouth they visited Goodrich Castle, where Herschel sketched with his camera lucida and Margaret gathered wildflowers. Usually a Wye tour ended at Chepstow near the

Goderich Castle Monmouth

FIGURE 8.2: Sir John Frederick William Herschel, *Goderick Castle, Monmouth.* Graphite drawing made with the aid of a camera lucida, unframed: 7 3/4 × 12 11/16 in. The J. Paul Getty Museum, Los Angeles, Gift of the Graham and Susan Nash Collection.

river's mouth, but Herschel and Margaret traveled farther, to the ruins of Tintern Abbey, and enjoyed the scenes there so much that they returned that evening. Moonlight on the ruins, Herschel recalled, was a view "which beggars description."[3]

Herschel was eager to take Margaret abroad for the second leg of their travels, but first they returned to Slough. In the midst of preparations for the next portion of their extended honeymoon, Herschel made time for a last-ditch attempt to convince the British government that the financial resources devoted to Charles Babbage's calculating engine had not been in vain and should be continued. Babbage's situation with his engine had not improved, and Herschel's friend felt trapped between recalcitrant artisans he had employed to construct the device and a government that had withdrawn its financial support. While Herschel and Margaret paused before leaving for Europe, Herschel agreed to meet at Babbage's house with some of his supporters to discuss what should be done next. The group decided that Herschel would request an audience with Arthur Wellesley (1769–1852), the Duke of Wellington, who had become prime minister the year before, to plead Babbage's case. The meeting with the new Tory prime minister, which took place the very next day, did not go well. Wellesley did not even give Herschel the chance to fully set out his case for Babbage. Herschel could only hope that a final answer from the government, even if it was no to additional funding, would release Babbage from his sense of obligation to continue work on the machine.[4]

Herschel's disappointing visit with the prime minister reinforced his discouragement regarding the state of science in Britain even as it confirmed his own prestige. In a letter to Whewell soon after the meeting, Herschel complained that Britain was "not a land where Science of a high order is held in honours." It was impossible not to lament the limited careers open to scientific talent. "For my own part," Herschel concluded, reaffirming his decision to withdraw from scientific society, "I have seen so much of this as to be heartily sick of the society of *Savans*." For the time being, Herschel was much more pleased with Margaret's company and eager to embark on the next phase of their travels, leaving behind for a time concerns over the state of British natural philosophy.[5]

The second portion of their trip would cover ground Herschel had traversed before. He enjoyed seeing the sights through Margaret's eyes and was constantly delighted with how their interests aligned. As they crossed the Channel and then passed from France into the Netherlands and on to the German kingdoms, his appreciation of her as a traveling companion grew, and he told his aunt Caroline that she was "a miracle of happiness . . . dropped upon me." The newlyweds were not alone for this portion of their trip, though. They took along Margaret's younger brother, Johnnie, who raced with Herschel to see who could make the most drawings of the sites they visited. In Brussels they rose at 5 a.m. to take sketches of the cathedral. Outside of Liege, Herschel, Margaret, and Johnnie toured the factories at Seraing, "the Soho of the Netherlands," where Margaret was impressed with the bellows and forge of the great foundry. There were also garden tours, picnics, a musical festival, and, at Margaret's insistence, readings aloud together from the life of Martin Luther. ("She loves me much," Herschel admitted, "but God more.") Their main complaint was the smell of tobacco, which Herschel said people in the Netherlands smoked or chewed endlessly. He teased Margaret about getting a cigar himself, which she said would make him "a filthy beast." Crossing into Prussia, they traveled by carriage along the Rhine and reached the German resort town of Baden-Baden by July.[6]

In Baden-Baden, the travelers felt that if their families had been with them, they would have been perfectly happy staying for months. They went about, Herschel reported, "sketching reading & hammering the hills, flute playing &c &c" and planned to continue into Switzerland to Berne. Their time was cut short, however, by an urgent letter from Grahame in London. Herschel's mother was ill, and Grahame urged them to return immediately. "Your dear mother is in a very precarious situation," Grahame scrawled in his nearly illegible hand. "We all think she is . . . on the brink of a precipice." Grahame vividly recalled the grief that had awaited Herschel's previous return from England. "I wish you to be near," Grahame insisted. "And I am convinced that such is her wish." He advised them to "come home directly."[7]

These warnings reached them just as they were about to board their coach to travel on to Switzerland, and Herschel and Margaret immediately cut their trip short. Herschel wrote a hurried letter to his mother: "Let me recommend you my dear beloved Parent to God's protecting care Whom I pray fervently to preserve you to us for many many years of health & happiness." Five minutes after receiving word of her condition, they were on the road home, hastening to Slough to find her largely recovered. Herschel was tremendously relieved. Concern for his mother's health had been constant since his father's death, and this scare ensured that for the remainder of her life Herschel would remain in England. For now, whatever scientific work Herschel did would be in his local context, from the domestic base he began building with Margaret at Slough.[8]

BROADER AUDIENCES

Fortunately for Herschel, transformations in the nature of publishing offered the opportunity to reach significant audiences without leaving home, and the connections he had made throughout the scientific world continued to serve him as he embarked on a project well suited for his new domestic context. By the 1820s, literacy rates in England were rising significantly, and a growing middle class was eager for books on the natural world and the practice of the new physical sciences. Reading was seen as a way of elevating one's social status, and an increasing number of encyclopedias and periodicals were being created to cater to this new audience. Organizations such as the Society for the Diffusion of Useful Knowledge, founded in London in 1826, published books and tracts aimed at developing the minds and habits of the popular classes through literary instruction.[9]

Herschel had already begun to establish himself as an author writing for an audience beyond that of scientific journals or members of learned societies. He had considered writing for encyclopedias as early as 1816, when, frustrated with tutoring pupils at Cambridge, he told Babbage he wanted "to get money *and reputation at the same time*" by writing articles. Money was no longer a primary motivation, but by now Herschel's reputation made him a sought-after author. In 1824 Herschel was approached by the editors of the *Encyclopedia Metropolitana* to contribute an article on optics, which became his treatise "Light." This was

followed by "Physical Astronomy," a technical and mathematical expo-
sition of astronomy that formed a bridge between the work of the Analytical Society and the Astronomical Society as Herschel illustrated the
importance of mathematical analysis to astronomical practice. (Though
written after "Light," "Physical Astronomy" was published first.) Next
came a similar state-of-the-field article on "Sound." These long pieces
were mathematically dense and well beyond the grasp of most general
readers. The chance to reach even wider audiences though came when
Herschel was approached to be part of a new project by an Irish mathematician named Dionysius Lardner (1793–1859) who was looking to
make his mark in the publishing world.[10]

Lardner, himself a well-known author and popularizer of science,
was preparing a new encyclopedia that would compete with the *Metropolitana*. The work, which Lardner was calling the *Cabinet Cyclopedia*, would offer more accessible, nonmathematical articles written for
a much wider audience. The project represented several publishing innovations. Published by Longmans, the largest firm in the encyclopedia trade, the *Cabinet Cyclopedia* would be unique in that its individual
volumes (which ultimately numbered 133) could be purchased as stand-
alone treatises or in "cabinets" of volumes on related topics. They would
also be priced for a wide readership; at just six shillings (about $36 today),
volumes were within reach of the growing middle-class reading public.
Writing for this new project would give Herschel the chance to reach a
large audience and present his ideas, specifically those on light and optics, in nontechnical format, a project he had already been contemplating.
Lardner had funding to enlist the highest talents for the project, he told
Herschel, and he agreed with him that a more popular treatment of the
concepts Herschel had presented in "Light" would be an ideal fit for the
Cyclopedia. Lardner offered two hundred pounds for the manuscript and
gave Herschel as much time as needed to compose it.[11]

Lardner's invitation came as Herschel was focusing on his astronomical surveys, but the project would give him a reason to return to writing and thinking about light, a topic for which he retained significant
passion and interest. He admitted that if Lardner had asked for a work
on any other topic, "I should without hesitation decline undertaking it,"
as he was still finishing "Sound," the third of his articles for the *Metropolitana*, and found writing it a considerable drain on his time. Even
Lardner's very generous terms would not have been incentive enough, he
explained, were it not that "the subject itself is one in which I take much
interest for its own sake and which I wish to become generally better
known & more popular than it is." The technical, highly mathematical
treatise he had already written, he told Lardner, had in fact been prepa-

ration for a more extensive work on the subject that he had hoped to one day write.[12]

Lardner was enthusiastic. Light was one of the most compelling and rapidly developing areas of physical science, and if Herschel would consider expanding the scope of his treatment to other aspects of physics as well, his contribution to the *Cyclopedia* "would constitute probably one of the most complete popular treaties on *physique* which has ever appeared." Lardner encouraged Herschel to think as broadly as possible in terms of scope, audience, and length. The book would be beautiful too: a treatise on light, "written in a popular style, conveying in ordinary language divested of mathematical formula the wonders of that fascinating department of physical science," could be "splendidly illustrated with coloured representations of the phenomena of the spectrum in all its modifications and varieties, of the rings by reflection and transmission," and "of the beautiful and various hues produced by polarized light." Lardner's vision for the book rapidly expanded. Such a volume, he said, would "diffuse a knowledge of and taste for the science most widely." Herschel was convinced, and he committed to preparing a popular treatise by the end of 1830.[13]

Unfortunately, Lardner's competition—the editors of the *Metropolitana*—were not as enthusiastic about the publisher's vision and protested when they learned of Herschel's plan. Their author writing a second article on light, they insisted, was "likely materially to injure the sale of their work." Who would want a highly technical treatment of light packed with equations when they could have an illustrated version by the same author focused on the concepts and containing vivid depictions of experiments? Their understandable objections put him, Herschel admitted to Lardner, "in an extremely painful predicament." He admitted he had perhaps been naive in assuming that by treating the subject in a different form and avoiding "everything in the proposed article which might place the two essays in rivalry" he could allay concerns of the *Metropolitana*'s editors. Herschel asked Lardner's advice. He still supported the *Cyclopedia* and wanted to be a part of it, but he did not know how.[14]

For Lardner, a major contribution from Herschel was key to the success of his publishing venture. Lardner's plan for the *Cyclopedia* depended on having well-known, well-regarded names contributing on their areas of expertise. No one but the popular Scottish author Sir Walter Scott (1771–1832), for instance, was good enough to write the volume on the history of Scotland. Not only would Herschel's name bring attention and prestige to the series; it would also attract other quality writers, especially for the volumes devoted to the sciences and mathematics. Lardner was prepared do whatever he needed to accommodate Herschel and keep him involved. He suggested Herschel and the Scottish natural philosopher David Brewster simply swap topics for the treatises they had agreed to write. He also remained extremely generous with terms: Herschel could take as long as he needed, and if the original fee for the work no longer seemed enough, he simply had to name his price. Ultimately, Herschel agreed to switch to the topic of astronomy, though he had also written a *Metropolitana* article on an aspect of that topic. He wrote Lardner explaining his choice only days before his wedding but admitted he could no longer meet the original deadline, "for the present year I should be effectually precluded by domestic arrangements." Herschel also asked that in advertisements for the series Lardner omit the "VPRS"—vice president of the Royal Society—from his name.[15]

Writing another long article, even a popular one, became much lower priority for Herschel as he started his new life with Margaret, especially since he would no longer be writing on his true scientific passion—light. Meanwhile, Babbage, fuming about lack of government support for his engine and ready to "beat to pieces the idol" he had created, had begun his own popular treatise, one that would highlight what he saw as the neglected state of science in Britain and would submit the continued failures and insults he believed he had received at the hands of the Royal Society to public view. The disappointing response Herschel received from the prime minister regarding funding had not helped matters, and the following winter Babbage began drafting a work he intended "to cut up the coterie" and "expose some of the abuses and there leave the thing to be corrected by public opinion." As Herschel prepared to write a text opening an aspect of contemporary science to a broad audience, Babbage was preparing one that would publicly air all his grievances and show that the pursuit and support of science in Britain was broken.[16]

Herschel was initially pleased with Babbage's idea, thinking it might do some good. "The state of science in England!" he responded after hearing Babbage's plan for his work. "Lord help us—we are a conceited nation and have our ignorance to learn. . . . It is not too late to correct that." Herschel thought that many enthusiasts of science in England (and in the Royal Society in particular) were little more than dilettantes, believing they were doing science but not understanding how backward their practice was compared to the rest of the Continent—while the true scientific authorities (like himself, Babbage, and other members of the dissolved Board of Longitude, for example) were largely ignored by policy makers and misunderstood by the public. Herschel hoped Babbage's work would be a corrective, promoting the methods of natural philosophy and emphasizing its importance to society as a public good.[17]

Herschel's own concerns regarding the state of science in England were similar to Babbage's, but he was more reserved about expressing them in print. His frustrations did, however, emerge at this time in an unlikely place: the final article he was finishing for the *Metropolitana*, his treatise on "Sound." There, buried in Herschel's discussion of the physiological basis for the perception of sound, Herschel inserted a footnote acknowledging that the majority of information for his article, as well as for "Light," came not from British but from European journals, specifically the French *Annales de chimie*. Herschel went on in the note to contrast the quality of scientific journals at home (with their "crude and undigested scientific matter") to those abroad. This was especially damning because as Royal Society secretary Herschel had played a role in accepting and commenting on papers for publication in the society's *Transactions*. He praised the French process of peer review, which, he said, meant meritorious scientific work "will immediately be reported on by a committee who will enter into all its meaning, understand it however profound, and not content with *merely* understanding it, pursue the trains of thought to which it leads, place its discoveries and principles in new and unexpected lights, and bring the whole of their knowledge of collateral subjects to bear upon it."[18] In a marked contrast to this competency of scientific review, Herschel had not long before received a plea from the leadership of the Royal Society to comment on a paper on crystallography submitted to the society that it seemed no one in London had the knowledge to review—and this despite the fact that the author had been awarded a Royal Society medal for his work on the topic![19]

The organization and strength of scientific publishing in Europe not only put Britain to shame; it offered opportunities to British natural philosophers that they might not even realize they had. Herschel explained, as the long footnote continued, that he was often surprised how quickly discoveries or research ignored at home was picked up by foreign scientific periodicals. British natural philosophers who felt unappreciated in their own country actually "have a larger audience, and a wider sympathy than they are perhaps aware of." Herschel spoke from the experience of having had his mathematical papers better known (and understood) on the Continent than in Britain. Any good experiment made in England, Herschel claimed, was immediately "repeated, verified, and commented on" in Germany and Italy. He wished the reciprocal was also true, and he lamented the fact that "whole branches of continental discovery are unstudied" and indeed unknown even by name. Even buried in a long encyclopedia article, the note was a severe indictment by one of Britain's scientific giants.

But there was more. As the note continued, Herschel's tone grew more despairing. "It is in vain," he wrote, "to conceal the melancholy truth. We are fast dropping behind. In Mathematics we have long since drawn the rein and given over a hopeless race." This admission was painful as well, as it meant the work of the Analytical Society, as well as the care and trouble Herschel had taken revising and publishing Spence's mathematical essays, had indeed been largely without effect in changing the landscape of British mathematics. Things were no better in chemistry, and Herschel provided a long list of specific areas of research he believed were being completely neglected. His remarks, he claimed, could be generalized to any other science. "The causes" for this decline, Herschel concluded the note, "are at once obvious and deep seated," but he refrained from going into detail.[20]

The long footnote, with its vague and ominous ending, was entirely out of keeping with the technical format of the rest of the article. It was a rare public expression of frustration and angst, coming at the conclusion of his years involved in the London scientific scene, but it might have remained an obscure screed in a long article if Babbage had not fixed on it for supporting the argument of his own treatise on the decline of science. By February 1830, when "Sound" was on its way to publication, Babbage's fortunes took a sudden turn for the better. The Duke of Wellington's government, despite the prime minister's initial hesitancy, renewed its interest in Babbage's engine and declared it "government property." This meant, so Herschel hoped, Babbage's financial difficulties would stabilize and put an end to the "awkward and embarrassing situation" that had caused Babbage so much anger and energy and allow him to finish "the great work." Despite this reversal, though, the only "great work" Babbage was focused on was airing his opinions and personal grievances in print.[21]

Babbage had received the promise of additional government support, but this did nothing, he felt, to restore years of wasted funds and effort he had already poured into the project. He continued work on his book and told Herschel that the arguments he was making had nothing to do with the status of his engine. He asked Herschel to send him proofs of the last page of his "Sound" article, now in press, hoping to cite Herschel's impassioned footnote in his own work. Herschel by this point may have regretted providing any additional fuel for Babbage's fire, and the pages he sent Babbage did not include the note. Undaunted, Babbage went to the printers in London and searched through their waste until he found it. The footnote, much to Herschel's chagrin, was reprinted in its entirety at the beginning of Babbage's book, *Reflections on the Decline of Science in England*, and inextricably linked Herschel to Babbage's arguments.[22]

Herschel was annoyed Babbage had pulled his footnote into the limelight and was even less pleased with Babbage's attitude now that governmental support for his engine—to the tune of 7,500 pounds—had rematerialized. "Who the deuce ever did anything worth naming without a sacrifice of some kind or other," he wondered. He told Babbage that Babbage should be ashamed of keeping up his "old growl" about time and money spent on a project that would bring him nothing but fame and contribute to the good of society. "If I were near you and could do it without hurting you," Herschel told his friend before he even had a chance to read the book, "I would give you a good slap in the face." He urged Babbage to burn the work rather than go ahead with publication.[23]

It was too late. "My book is written," an unrepentant Babbage informed Herschel, "and goes to press perhaps tomorrow." He admitted the work would make him enemies and that the government would likely be annoyed. But he hoped "to teach even chartered and eminent bodies a lesson," referring specifically to the Royal Society. He would "make them writhe" with the ax of public opinion. In short, Babbage maintained, his volume would "be a receipt in full for the amount of injury I have received"—and in terms of its immediate effect, he was correct. Though the *Decline of Science* was not a financial success, selling fewer than a thousand copies and not making it beyond a first edition, the response to the book was immediate and polarizing. Many readers sympathized with the author and appreciated Babbage calling out the failings of the Royal Society. But others felt that for the recipient of such significant financial support, the book was ill timed and ungracious. In his work, Babbage went so far as to single out specific individuals he felt had slighted him or deserved censure for their scientific claims and then expressed shock when they were offended. In short, the *Decline of Science* added to division and strife in the London scientific scene. When it came to reforming natural philosophy, Babbage had taken the same tack that had so frustrated Herschel in their student days at Cambridge: pouring fuel on the fire with his personal pique.[24]

At the time of the *Decline*'s publication, the only drama Herschel was interested in navigating was the domestic drama of preparing a home for a new baby. Margaret was pregnant with their first child, who—if it turned out to be a girl—they decided would be named Caroline, after Herschel's aunt. Even as Babbage prepared his *Decline of Science* salvo, the Herschel home at Slough was undergoing major renovations in the transition from residence of a bachelor astronomer and his aging mother to the home of a growing family. Herschel first labored to get his "chemical room in order." After that, the house became a mess of replastering, painting, and putting down new floors. Things were in such disarray, Herschel confided to a friend, that "a *Pianoforte* was searched for and could not be found in the melee." Margaret, perhaps wisely, remained in London with her mother and siblings while the work was under way, with Herschel sending her frequent updates. By the middle of January 1830, things at Slough were prepared, and Margaret moved in.[25]

In the midst of these domestic transitions, Lardner approached Herschel again, this time with another opportunity—one that would give Herschel the chance to respond to the tumult raised by Babbage's book. Lardner had decided to divide his *Cabinet Cyclopedia* into multiple "great divisions," and his latest brainstorm was that each division would be prefaced with a book-length essay written by one of "the highest talents of the age." The idea was not wholly original. There was a tradition of extended "introductory treatises" prefacing works in natural philosophy, especially in France, in the late 1700s and early 1800s. Cuvier and Laplace, for instance, had both written long introductory prefaces to their own major works.[26]

Lardner thought Herschel was the ideal candidate for writing a prefatory essay to the division on natural philosophy; Herschel would be "more effectually instrumental probably than any one now living." Hearkening back to Lardner's original hope that Herschel's treatise on light might expand into a treatment of all branches of physics, the subdivision of the *Cyclopedia* for which Lardner wanted Herschel to write the introduction would include physics as well as mathematics and chemistry—broadening the scope of Herschel's potential work to all of natural philosophy. Lardner left the length of the introductory essay to Herschel but hoped it might be two-thirds the length of his projected volume on astronomy. He offered 250 pounds for the work and asked Herschel to write the essay *before* his treatise on astronomy. Herschel did not deliberate long. Within two weeks, he responded that "after the best consideration which I have been able to give" he felt he could write the prefatory essay and have it delivered in about six months' time.[27]

Herschel, as Lardner recognized, was ideally positioned to offer commentary on the practice of natural philosophy, both as a practitioner of all the branches of science that would be discussed and as a stymied reformer to the practice of science itself. The work would be his chance to formalize his thoughts on natural philosophy and the role it could play both morally and socially, synthesize what he had learned in his travels, and offer a cogent response to Babbage's claims of decline on the one hand and the conservative voices in the Royal Society on the other. The book he set about to write would be the climax of his own scientific self-fashioning, drawing on his experiments, correspondence, experiences, and ideals. He could not know it at the time, but the months he spent drafting what would become his *Preliminary Discourse on the Study of Natural Philosophy* culminated in more than the immensely successful articulation of his view of science: it coincided with the most dramatic moment yet for scientific reform in London.

WRITING THE *DISCOURSE*

As Herschel began drafting his prefatory essay, he was in the midst of two other writing projects. Besides responding to Babbage's *Decline of Science*, published in May of 1830, Herschel was corresponding extensively with mathematician and author Mary Somerville (1780–1872) as Somerville worked to complete a nonmathematical version of the first volumes of Laplace's *Mécanique celeste* for a general audience. This was a project very similar to Lardner's goal for the *Cabinet Cyclopedia* of presenting scientific concepts for wider audiences, and Herschel thus began his own work while carefully reading and commenting on Somerville's drafts of what would ultimately become her *Mechanism of the Heavens*, published in 1831. Somerville's synthesis of the concepts in Laplace's magisterial treatment of celestial mechanics provided an example of writing that expressed abstract mathematical analysis in straightforward prose. What Somerville was attempting for the solar system, Herschel hoped to accomplish for the practice of science itself.

Herschel was also at work at this time on a short summary of the mathematical and scientific accomplishments of Thomas Young, the British natural philosopher who had supervised the *Nautical Almanac* and done so much to develop the wave theory of light. Young had died the year before, and Herschel had agreed to contribute to a biography being prepared by Young's friend Hudson Gurney (1775–1864), a member of Parliament and Royal Society fellow. The challenge for Herschel was that Young had arrived at his important physical insights without what Herschel saw as the proper methods of analysis. Considering Young's scientific career forced Herschel to revisit the question of how scientific discoveries were made.[28]

This topic of the means of discovery went all the way back to Herschel's earliest work with Babbage on analysis, when the two had argued that accurate mathematical representation was key to clear thought and could even be a means of mechanizing the process of discovery. Herschel had pursued these ideas in his experiments on optics; Babbage had turned to actualizing this in the creation of his difference engine. Herschel now returned to the topic as he considered Young's work, even as Babbage maintained that it would be impossible to write any account of Young and his career that did not "mention the evil influence he has had on the science of the country." A treatment of Young that failed to mention his failings, Babbage warned Herschel, would "be most unjust and most injurious to science." Though Herschel disagreed, trying to compose a summary of Young's career that explained his method of insight plagued Herschel over the next few months while he articulated his own views on scientific discovery.[29]

Any serious writing was interrupted, at least briefly, by the birth of Margaret and Herschel's first daughter, Caroline, born 31 March 1830. The interruption did not slow Herschel's thoughts on the topic, however, as the very next day, riding back from London where he had conducted some banking and put in place plans for Caroline's christening, Herschel had the idea for a work on logic, "or rather on the Principles of Reasoning." This thought became the genesis of a major portion of what Herschel was already calling his "discourse on science." Rather than simply an account of the practice of natural philosophy or an apology for the scientific life, his discourse would attempt to answer the question that haunted him about Young's work: not just how science was done but how one *thought scientifically*. This would make Herschel's book more than simply an appreciation of science; the *Discourse* would become an invitation to develop the mind of a natural philosopher. Even with this inspiration though, the amount of writing he had before him, coupled with long evenings at the telescope and work during the day grinding a new set of mirrors for the twenty-foot, left him overwhelmed. "I am at this moment," he wrote to his mother-in-law not long after Caroline was born, "*physically incompetent* to get through the mass of writing which at this time overwhelms me."[30]

By the end of May, Herschel had read Babbage's book, where he found the words from his footnote used as one of Babbage's prime testimonials on the decline of science in England. Babbage's views and the response Herschel could provide with his own book were no doubt a central point of conversation the next month, when two of Herschel's closest friends, Richard Jones and William Whewell, came to stay at Slough for a weeklong visit. Whewell was by this time chair of mineralogy at Cambridge, and Jones was working on his own book, an inductive treatment of economic theory. (Herschel had contributed to this project by sending books on Dutch and Portuguese commerce home to Jones from his trips abroad.) Both of his friends were, like Herschel, dismayed by the vindictive tone of Babbage's treatise. Though Jones and Whewell agreed the Royal Society needed reform, Whewell wanted nothing to do with "the Babbagian sect of spewers or railers," and Jones predicted Babbage would only injure himself by his tirades. Their discussions no doubt harkened back to their Cambridge debates and helped Herschel sharpen the arguments he was making in his book—arguments that would impel Whewell to his own extensive writings on the philosophy of science.[31]

After Jones and Whewell departed, Herschel finalized his account of Young's work and in so doing provided an encapsulation of the process of inductive discovery he was working out in his own book. Young's most important scientific contribution, according to Herschel, was his generalization of the principle of interference from vibrational motion (such as waves on water) to other cases—initially sound but eventually and most importantly light itself. "In science," Herschel wrote in his account of Young's work, "he who generalizes, discovers." Yet it was not enough to simply generalize a principle like interference from analogous physical situations; it was also necessary "to convert inventions of simple generalization into valid and useful discoveries." These discoveries were made by "inverting the ladder of induction" to apply the new, generalized principle to specific cases—most often through experimental confirmation. Without this, "generalizations are matter of pure speculation, and, in fact, nothing more than mere forms of words." In other words, arriving at a generalized law of nature (such as the interference of light) was only half of the process; that law then needed to be applied in new and surprising ways to other situations. Young had been able to take this additional step by testing the interference of light using various materials on hand: "a scrap of card, a hair, and a candle; a few scratches on a bit of glass held in the sun and turned slowly round; a piece of paper, a pinhole, and a closed window shutter." These experiments proved interference and thus the wave nature of light, a law "so important as at once to change the face of optical science."[32]

As Herschel was articulating his views on scientific discovery, he was also still pondering how the practice of natural philosophy was related to virtue. In July Herschel found a passage from Cicero's treatise on ethics, *De Officiis*, that he told Margaret "will suit me to a *t* for a motto to my 'discourse on science.'" In the passage, which was printed in the *Discourse* after the table of contents, Cicero states that "before all other things" humanity is distinguished by the "pursuit and investigation of TRUTH." People "delight to see, hear, and to communicate" truth, which is "necessary to the good conduct and happiness of our lives." Herschel was using Cicero as more than an argument that natural philosophy was beneficial because it *revealed* truth; rather, natural philosophy had a virtuous moral and political purpose. Those who understood science, Herschel would go on to spell out in the *Discourse*, were less likely "to yield obedience to any orders but such as are at once just, lawful, and founded on utility." Herschel was making it clear from the very beginning of his text that natural philosophy was the basis for a virtuous society.[33]

This connection between natural philosophy and virtue catalyzed Herschel's writing. The Cicero quote, he told Margaret, "has rather electrified me so that I am all in a ferment—going about the garden looking how the trees will grow and threatening what fiery things I shall (one day) write." At the same time, Herschel was also electrified by the progress of his astronomical surveys. He had completed polishing two new telescope mirrors, which he had been working on for several months. These were the first he had prepared without his father's supervision, and he was quite pleased with their quality. The mirrors made his sweeps the clearest they had ever been, allowing Herschel to see details and detect objects with his telescope that he believed "no other telescopes that I have yet heard will touch." This wonder that he felt at the new objects swimming into his eyepiece (including Uranus, which he accidentally "rediscovered" on his first sweep with the new mirrors) infused his prose as well.[34]

Alongside coaching Somerville in writing for a popular audience, wrestling with ideas of discovery in Young's work, and experiencing astronomical discovery in his own research, as Herschel neared completion of the first draft of his *Discourse* a family tragedy took place that provided a poignant counterpoint to the joy of their new child and Herschel's growing passion for his project. Margaret's younger sister, Isabella, had been sick for some time, but in July of this summer she became "alarmingly ill." Margaret left baby Caroline in Herschel's care to stay with her mother and sister in London. By the end of the month, it became clear that the sixteen-year-old Isabella, Margaret's only sister, would not survive. She died the next month, on 26 August, her brother's birthday. Echoes of the tragedy made their way into Herschel's manuscript as Herschel ensured that his view of natural philosophy would not undermine the consolation the Stewarts found in their faith. True contemplation of science, he made clear in the first pages of the *Discourse*, does not lead to despair or conflict with religion. Rather, natural philosophy will for its practitioners "first encourage a hope, and by degrees acknowledge an assurance, that . . . intellectual existence will not terminate with the dissolution" of the body. The natural philosopher could be confident that they would "drink deep at that fountain of beneficent wisdom" in "a future state of being." Isabella was gone, but the treatment of science Herschel was composing offered comfort to Margaret and her family and reassurance to any reader concerned about the influence of science on religion.[35]

As September brought to an end an eventful and tragic summer, Herschel had completed the initial portions of his book, and Lardner was sending back from the printers the first proofs. The editor was quite pleased with the work, but as the *Discourse* neared publication a significant question remained regarding Herschel's relationship with the Royal Society: what letters would appear after his name on the title page? As Herschel made the *Discourse* his appeal for the virtuous life of a natural philosopher, how should he acknowledge his relationship to the institution that had come to many reformers to symbolize much of what was wrong with natural philosophy in Britain? Lardner was unsure what Herschel's intentions were regarding titles and honorifics but was convinced whatever Herschel decided would have repercussions. "I do not know," Lardner wrote to Herschel, "whether it is by intention or accident" that Herschel had omitted the usual "F.R.S." (fellow of the Royal Society) after his name in the submitted manuscript. "The omission of this at present immediately after the appearance of our common and highly esteemed friend Babbage's Pamphlet," Lardner pointed out, referring to the *Decline of Science*, "will no doubt be considered as an intentional slight upon the society." In the aftermath of Babbage's attack on the Royal Society, whatever Herschel did would be a statement—either of support or censure.[36]

Herschel responded that he did not intend to slight the Royal Society or make a protest but that if he included "F.R.S." after his name, he would in fairness need to include titles from the "twenty or thirty academies and institutions" of which he was also a part. He could "not draw a line," he told Lardner, between the Royal Society and these others. Gone were the days, at least in Herschel's mind, when the Royal Society dominated the scientific landscape. For him, it was now simply one scientific organization among many. Readers of Herschel's work would thus see the best-known scientific name of the time conspicuously bereft of *any* scientific title, including that of being a fellow of the Royal Society. Just as he had maintained to Whewell and his colleagues when refusing to put his name forward for Cambridge positions, Herschel wanted it clear that he was writing his *Discourse* as an independent natural philosopher, not a member of any particular institution or learned society. By refusing to acknowledge affiliations, Herschel supported the claims of the *Discourse* that science was truly democratic, open to anyone willing to learn to think scientifically. Ironically though, within a month of telling Lardner he would forgo acknowledging any formal ties to the Royal Society, Herschel was proposed to lead the very society he had in some sense disavowed.[37]

FINAL REFORM

Even as Herschel decided to forgo in print his association with the Royal Society, events were drawing him out of his retreat at Slough and into a culminating confrontation with the conservative parties at the center of the London scientific community. In October, as Herschel was finishing revisions to the *Discourse*, rumors appeared in a London paper that Herschel was being proposed as a candidate for presidency of the Royal Society. Herschel was definitely uninterested. He had worked too long to distance himself from the rivalries and controversies of the London scientific community to be drawn back into them. "I love science too well," he wrote to Babbage, "to be easily induced to throw away the small part of one lifetime I have to bestow on it on the affairs of a public body which has proved to me ever since I became connected with it a continued source of disquiet and annoyance." He had found a rewarding productivity as a writer and astronomer holding himself aloof from the institutions of science, especially in the wake of the controversies stirred up by Babbage's book.[38]

Herschel's determination to remain uninvolved was only bolstered when he received a letter from William Henry Fitton (1780–1861), another scientific reformer and lately president of the Geological Society of London. Davies Gilbert, the Royal Society's current president, was planning to resign, and Fitton gave evidence that Gilbert was in correspondence with Augustus Frederick (1773–1843), Duke of Sussex and younger brother to the king, about taking the position. Here was another blatant example of one of Herschel's greatest frustrations with the society: private dealings in which Gilbert and others treated leadership positions as matters of personal privilege. Others agreed. An article in the *Times* bemoaned the members of the Royal Society treated in such a "disrespectful manner, like a herd of swine, [sold] to the best bidder, without even being consulted on the occasion." Another article drew a parallel between dealings with the Royal Society and calls to address parliamentary corruption: Gilbert's negotiations were compared to "borough mongering," the unpopular practice of buying or selling parliamentary seats.[39]

Herschel was furious with these machinations, telling Fitton he had trouble finding words to express his disgust. Gilbert's actions, he felt, showed the president had neither capacity nor dignity to hold his current position. Herschel deplored the assumption that the presidency could be passed on to the duke as though it were a "rotten borough," another clear link to the broader political context. Just as rotten boroughs allowed a patron or landowner to gain outsized political influence by swaying or purchasing votes of a borough with a very small electorate (a practice the Reform Act would do away with in 1832), Herschel saw Gilbert overextending his own influence over the fellows of the society. To treat the society like a rotten borough undermined the scientific meritocracy the society should be cultivating. Fitton's revelation confirmed Herschel's determination to have nothing more to do with the organization. There was no possibility, he said, of anyone "of *moderate* spirit and independence . . . taking part in the administration of the Society's affairs under the proposed new regime."[40]

On the other hand, the reforming party in the Royal Society could not simply let Gilbert hand the presidency to the duke. Herschel's suggestion of a potential candidate to stand against him revealed the remarkable extent of his reforming ideals. The best candidate for the Royal Society presidency, Herschel felt, was Francis Baily, retired stockbroker and leader in the Astronomical Society who had made a name for himself by his extensive star catalogs and systematic calculations reducing astronomical data. To Herschel, Baily exemplified the new kind of scientific gentleman revitalizing British science: someone who united the values of business with natural philosophy and who had gained credibility through mathematical and scientific merit, not hereditary position or status. That Herschel could propose Baily for presidency of the Royal Society against His Royal Highness, the Duke of Sussex, showed his radical idealism regarding scientific society: in Herschel's view of science, Baily's star catalogs could stand against all the aristocratic privilege of the duke.

A Baily presidency would have indeed represented radical reform. Gilbert and his conservative allies had long argued that aristocratic qualities were essential for a Royal Society president, but none of these—social graces, connection with government, the ability to host gala events—were qualities Herschel gave in support of Baily. Baily, according to Herschel, was best suited to be Royal Society president because of "his singularly regular, methodical habit of business and his disposition to avail himself of talent wherever he can find it." In other words, in the society that Herschel and the other reformers envisioned, it was more important to have an efficient scientific manager than an aristocratic patron as president. It was this managerial skill along with "his high scientific qualifications and his real love of Science for its own sake" that made Baily the better candidate.

In contrast, the social status of the Duke of Sussex actually undercut his personal qualifications. "His rank is too high to *permit* his exercising the most important part of the duties of the office of President—that of mixing familiarly and intimately with his *brothers scavans*." There would be difficulty "*working* in *subordination* to an individual whose requests [are] habitually considered as commands and his wishes guides of conduct." For Herschel the virtues of science were best exhibited when pursued without ulterior motives. "The *drudgery* of one who really takes a part such as modern science requires"—a drudgery Herschel knew from personal experience both in the bureaucratic functioning of the society and the meticulous observations and calculations of his own projects—"is *most severe* and can only be borne when its *motive* cannot be mistaken." If Sussex was requesting the work, for example, the pure pursuit of science became difficult to distinguish from obligation to royalty.

Babbage's book complicated the matter further. Because Babbage had brought attention to rewards and financial compensation in the practice of science, his treatment had the effect of "systematically lowering the motives of all scientific men," making it more important than ever for those with "the dignity of science at heart" to avoid situations where their actions could be misinterpreted. This was exactly what would happen with a powerful patron like the duke directing the society. Misinterpretation of motives was corrosive to Herschel's scientific ideal, and unfortunately Babbage's book itself was "a specimen of the extent to which such a misinterpretation may be carried." For the purity of science and its pursuit, the duke's candidacy had to be opposed. Someone needed to run for the reforming party, and Francis Baily was Herschel's candidate. Herschel told his friends in London he was committed to working with Baily and supporting him if elected.[41]

When it came to his own path, Herschel still felt the best way he could help the cause of reforming science was to continue in his retirement at Slough simply *doing* science. In this manner, he could honestly report that he was working "while nine tenths of the Royal Society are sleeping," his own efforts sustained by "the idea that the little I *can* do for science is at least all clear gain and is not liable to be neutralized or turned into evil by the passions and disputes of others." By withdrawing from the London scientific scene, Herschel retained full control over his own work. He could set his own priorities and was confident his results would not be overturned by changing governments or goals, as the efforts he had poured into the Board of Longitude were. "Why then," he wondered, "should I quit a field where I know that my efforts be they more or less at least have their full effect" and return to obligations "which I may never afterwards have power to free myself from"?[42]

Yet the threat that the duke's candidacy represented was enough to make Herschel revise his determination to withdraw completely. Gilbert's handling of the question of presidential succession violated Herschel's sense of justice, and he felt obligated to return to London to consult with fellow reformers about the best course of action in response. At a meeting at Fitton's house in late October, a new solution emerged: Herschel's colleagues urged him to take seriously the suggestion that he stand against the king's brother for the presidency. Suddenly, the pressure Herschel was under to complete his *Discourse* became overshadowed by a larger concern: whether or not to return to the scientific limelight in the most dramatic way possible—running for Royal Society president against a member of the royal family. It was one thing to suggest Baily be put forward; it was another to agree to it himself. To make such a public election contested was a serious matter. In the previous presidential election, Robert Peel had refused to run because Gilbert could not assure him his election would be uncontested. Babbage, likewise, had not wanted his name to go forward for secretary for the same reason. If someone of Herschel's stature ran, not only would the election be contested, but it would become a contest that would split the society between conservatives and reformers. It would also be a public rebuke to the duke, indicating a lack of agreement within the society that he was the best person for the job. For someone as averse to personal controversy as Herschel, this was an incredibly unwelcome situation. Moreover, as he confided to Margaret, his book was advertised to appear on the first of December, just over a month away, and he had still not finished it.[43]

At the end of November, just before his book was due to be delivered to Lardner, there was another advertisement, this time in the London *Times* and this time making Herschel's candidacy more than a suggestion. First published in the 25 November issue, it reappeared again on 29 and 30 November. Ultimately bearing the signatures of eighty members of the Royal Society—including reformers Babbage, Baily, and Fitton; Cambridge colleagues William Whewell and Adam Sedgwick; instrument makers George Dollond and Edward Troughton—the advertisement was a public request by his colleagues for Herschel to run for the office. His appointment, it stated, "would be peculiarly acceptable to men of science in this and foreign countries," and the undersigned intended to give him their vote. The notice, Herschel admitted, placed him in a "very difficult predicament." Either he refused "a duty to which I am called by a large number of respectable men, many of them my personal friends," or he incurred "the *most distressing* interruption . . . to those objects which I have made up my mind to regard as the main business of my scientific life," namely his astronomical pursuits and study of light.[44]

It was duty to his scientific colleagues that convinced him. Herschel, Babbage, South, Fitton, Baily, and others had campaigned for reform in the Royal Society for years, and Herschel felt he could not abandon that effort now, despite the fact that he had no desire for the responsibilities the position would entail. Herschel told Babbage he would agree to let his name go forward but that if he won the election, he would hold the position for only a year. A year with a reform-minded president, Herschel felt, would be enough to make genuine and lasting change, carrying out the reforms that had been initiated and then shelved by Gilbert. "I do not *desire* the Presidency," he emphasized, but "if placed in the Chair I will sit there one year & work." Herschel planned to name Babbage to the council and nominate members of the reforming party to positions of leadership. The Royal Society would be restructured so it functioned more like the Astronomical Society or the Geological Society, those upstart institutions that had originally threatened the hegemony and eroded the exclusive influence of the Royal Society since their formation.[45]

Herschel confirmed his decision to stand for the election by notifying the Royal Society of his intention on 26 November. He emphasized that he made the decision out of duty and not ambition, but he made his reforming intentions clear. "I see clearly that the R.S. is beset with difficulties," he explained, and he would do his part if elected to implement membership and policy reforms. The presidency would be divested of its autocratic tendencies, and Herschel would promote a scientific meritocracy and the policies of efficiency and transparency that Babbage had argued for in his book. In the days running up to the election, Herschel hoped his supporters would remain dignified and that "no ill-judged step will compromise the motives of any one concerned." He was concerned, as usual, about the unpredictable antics of hot-tempered colleagues like Babbage and South.[46]

Herschel did not have long to wait in suspense. The election took place on 30 November, giving many of the fellows who supported Herschel barely enough time to make the trip to London and cast their votes. As Herschel anticipated, his candidacy was polarizing and contentious, with newspapers remarking on the impropriety of the son of William Herschel, who had been patronized by King George III, opposing an election of George's son. A writer to the *Times* maintained that Herschel's candidacy showed ingratitude, claiming (incorrectly) that Herschel's fortune was in large part due to the generosity of the duke's father. A vote for the duke represented the status quo, maintaining the influence and prestige of aristocratic leadership in the society. A vote for Herschel represented reform, moving the society toward the businesslike efficiency of the Astronomical Society. Unfortunately for Herschel and the reformers, their party underestimated the strength of the conservative block and support for the duke while overestimating the number of those who supported reform. (Babbage even told some of Herschel's supporters not to bother to make the trip because his victory was assured.) Herschel lost the election by only eight votes, 119 to 111. The conservatives had retained power in the face of the most serious attempt yet by reformers to take control. The Royal Society preserved the aristocratic tenor that Gilbert and his allies believed was essential for its survival and influence. It appeared as though scientific reform had again been defeated.[47]

But even as he lost at the polls, Herschel was planting the seeds of the ultimate success of his broader reform of science. On 17 December Herschel submitted the final revisions of his *Preliminary Discourse* to Lardner. Herschel's reform of science would indeed come, but it would not be through his reluctant leadership of the Royal Society. Instead, it would be realized by his words on the page and the democratic vision of science he created there, a vision embraced by the next generation of natural philosophers—the first scientists.

CHAPTER 9

THE INVENTION
OF SCIENCE

There is no doubt in my mind that its publication will create a
greater sensation in these countries and elsewhere than any scientific
work which has ever appeared.

—DIONYSIUS LARDNER TO JOHN HERSCHEL, 2 OCTOBER 1830

The year was 1836, and Charles Darwin was on his way home. The
young naturalist had been aboard the *Beagle* for more than four years,
providing companionship to the ship's philosophically minded captain,
Robert FitzRoy (1805–1865), and gathering observations on the geology
and natural history of regions from the Galapagos Islands to Tierra del
Fuego at the southern reaches of South America. These observations
would provide the basis for his theory on natural selection outlined in
his *Origin of Species*. Darwin's biological revolution was still decades in
the future, however, as the *Beagle* coasted into the waters of Simonstown
Harbor in the British colony of the Cape of Good Hope in South Africa. His eyes were drawn northward toward Cape Town and the slopes
of Table Mountain, where he hoped to meet for the first time the person who was in many ways his scientific idol and role model, Sir John
Herschel.

Herschel had come to Cape Town, on the fringes of the British
empire, to complete the task he had set for himself of extending his
father's observations to the entire heavens. Herschel had revisited his
father's observations of the northern sky, resulting in a catalog published
just before his departure from England that included five hundred new
nebulae and formed a nearly complete record of all sidereal objects vis-

ible from the northern hemisphere. The catalog, along with his second contribution to Lardner's *Cabinet Cyclopedia*, the *Treatise on Astronomy*, brought Herschel even wider public and scientific esteem. If anything, Herschel's prestige in the scientific community had only increased after his defeat by the Duke of Sussex. His position as an independent savant working outside any scientific institution gave him authority as a spokesman for the life of a natural philosopher. These perceived virtues of self-sufficiency, humility, and reserve were magnified by his retreat to the Cape of Good Hope, especially when he declined multiple offers from friends and colleagues within the Admiralty for transport to the colony at the government's expense. By his withdrawal to the edge of empire, Herschel removed himself from the London scientific community about as far as was possible. Herschel, by the time of Darwin's arrival, had established himself at the Cape as a stargazer at the world's edge, living the life he had advocated in the *Preliminary Discourse on the Study of Natural Philosophy*.[1]

Herschel's appeal for the scientific life had been one of the driving influences on Darwin as a student at Cambridge trying to decide whether to pursue a career in the church or natural philosophy. "Few books," Darwin reflected later in reference to the *Discourse*, "ever made such an impact on me." Not only did Herschel's book offer a systematic account of scientific methodology—providing a roadmap for theory formation that Darwin would follow as he created his own theory on the origin of species—it also laid out the virtues of the scientific life, virtues upon which Darwin would model his own vocation. "If you have not read Herschel in Lardner's *Cyclo*—," Darwin advised a close friend, "read it directly."[2]

Darwin, though one of the most well-known examples of the *Discourse*'s influence, was in no way unique. Immediately upon its publication, the *Discourse* had become a sensation, selling thousands of copies. Reviews held the book up as the first systematic treatment of scientific method. When Darwin's colleague FitzRoy published a poorly executed volume on geology, a reviewer recommended the captain read Herschel's *Discourse* "before he ventures again in the same direction." More than a successful summary of science, the *Discourse* was a "philosophy of physical science," providing "the principles on which its structure rests, the maxims by which its researches have been and must be successfully conducted." It would be a book, according to the review written by Herschel's close friend William Whewell, "in the hands of all English readers" affected by the march of progress. Michael Faraday, the well-known authority on electromagnetism, told Herschel that the work had made him "a better reasoner and even experimenter" and moreover "has altogether heightened my character." Historian Sir James Mackintosh, in his own contribution to Lardner's *Cyclopedia*, maintained that Herschel's book was "the finest work of philosophical genius" of the age.[3]

Again and again, reviewers of the work noted that Herschel was the only person who could have written the *Discourse*. Herschel was "eminently qualified" to write the work because of his wide grasp of the branches of natural science. Herschel had "abilities and accomplishments" possessed by few others, giving him both the capacity to pursue original research and to generalize a comprehensive view of science with full scope of "its value and dignity." The work, which was the "first attempt since Bacon to deliver a connected body of rules on philosophizing," was the only such attempt by someone whose own work in natural philosophy, mathematics, and experimentation gave "examples of their successful application." More than a treatise on method, the *Discourse* was "one of the finest essays on the moral conduct of the intellect which has ever been produced."[4]

This last sentiment was echoed in the *Wesleyan Methodist Magazine*, which, with a much higher circulation than literary review publications, provides a perspective on how Herschel's book was perceived in middle-class and evangelical circles. The reviewer calls the work an apology for scientific pursuit that stands out for its accessibility. The *Discourse* contained "no mysticism; no purposed and pedantic obscurity; no elaborate trappings of scientific pageantry; no wrapping-up the stately forms of scientific truth in technical swaddling-clothes." There is nothing in the book to indicate Herschel is speaking only to the initiated; rather, he makes the methods of science clear to all. The review, after walking the reader through a careful discussion of the methods of reasoning that Herschel outlines, linked this methodological approach to Herschel's pious conclusions. His work lacked the "spirit of arrogant and presumptuous skepticism" that formed "the bane of all speculative philosophy." The *Times* agreed, remarking that Herschel's work rendered the practice of science "intelligible to the general reader," and brought it down "from that region of clouds and mists in which it is often concealed from the eyes of ordinary men."[5]

As reviews of the work emphasized, before Herschel composed his appeal for the scientific life and the moral value of science, he had embodied it in his own person and career. All the considerations, themes, and experiences from Herschel's life and career flowed into the writing of the *Discourse*, the book that became Herschel's vehicle of scientific reform. From his earliest days at Cambridge, Herschel had strived to reform the role of mathematics in reason itself and eventually as applied to the natural sciences. Though the work of the Analytical Society did not bear initial fruit, as curricular changes finally began at the university through the labors of Whewell, George Peacock, and others, Herschel used the *Discourse* to articulate the importance of mathematical analysis in natural philosophy. If his readers could not themselves practice the advanced mathematics science required, at least they would understand the essential role modern analysis played in scientific theory. The *Discourse* allowed Herschel to complete the transplantation of Continental analysis to British soil.

Herschel's close connections with natural philosophers across Europe and his frequent trips abroad had given him a view of science that was international in scope, allowing him to focus on the traits of scientific thought common to all national and institutional contexts. His extensive travels also gave him firsthand experience applying his approach of measurement and quantification to mountaintops and volcanic peaks as well as to the starry skies of England. Herschel's journeys provided a background to the treatise on natural philosophy he produced. It was through travel that Herschel forged his view of science, a vision that combined the power of the discipline to unify human thought of the physical world—from atoms to mountains to stars—with the personal virtue of the practitioner.

There were both personal and market motivations at play in the creation of the *Discourse* as well. The *Discourse* was an outgrowth of Herschel's evolving religious and theological beliefs. Though the book fit squarely within mainstream English religious thought, Herschel's ideas regarding faith were strengthened and developed through his relationship with the Stewarts and his marriage to Margaret, increasing Herschel's sense of the importance of stability in society at large as well as a domestic stability that mirrored and supported this in the home. The scientific output Herschel produced as well as the book he wrote celebrating this life arose from this environment. At the same time, the *Discourse* was a direct product of forces in the publishing world and the ideas of Dionysius Lardner to compete in the crowded market of encyclopedias and books for popular audiences.

Herschel's conflicts and frustrations with the Royal Society in particular and the London scientific community more generally provided an immediate background to the *Discourse*, as Herschel ultimately gave up on reforming the institutional structures of natural philosophy even as his credibility and influence within them grew. Instead of engaging in direct and public controversies like Babbage, Herschel chose to withdraw further into his own scientific practice. Paradoxically, this gave him even more credibility to offer his critique of the autocratic nature of the Royal Society. Herschel had decided that the true path to reform was reforming natural philosophy itself, not its institutions, by making its practice open to all—ultimately creating a broader, more democratic vision of science. The *Discourse* was a reforming text, and it helped transform natural philosophy into the practice of modern science.

REFORMING NATURAL PHILOSOPHY

Herschel's grand vision of natural philosophy gave it many of the modern connotations of today's science and unfolded in the *Discourse* in a three-part structure. First, Herschel laid out the practical and moral importance of natural philosophy, emphasizing its role in the material benefit of society as well as in cultivating virtue among those who applied themselves to its methods. In the second portion, which was seen by many contemporaries as the most philosophically significant section of the book, he outlined the methods of scientific reasoning that allowed natural philosophy to bear this practical and moral fruit, providing rules of reasoning from observation that would set off epistemological debates for later writers like John Stuart Mill and William Whewell. Laying out these methods of scientific reasoning set the stage for the third and final portion of the *Discourse*, in which the results of scientific reasoning are arranged to provide an overview of the contemporary fields of the physical sciences.

Herschel began the first of these three portions by insisting that the proper use of human reason has a moral significance. In the first instance, reason leads to an understanding of a world "disposed with order and design." This results in "the conception of a Power and Intelligence" superior to all human intellect with power "adequate to the production and maintenance" of the physical world. At the same time, the limits of human knowledge give rise to a hope that those deficiencies will one day be satisfied. From the very beginning of his *Discourse*, Herschel points toward a creator and the hope of eternal life, refuting a view that science would lead to doubts on the immortality of the soul or a distain for "revealed religion." Instead, reason provides a path between "enthusiasm and self-deception" on the one hand and irreligion on the other by its encouragement of belief in a life beyond the present state. "The character of a true philosopher is to hope all things not impossible, and believe all things not unreasonable." Humility and hope are the first of Herschel's scientific virtues, and they point toward a stabilizing faith that eschews both evangelical excesses and secular doubt.[6]

The position that natural philosophy leads to faith in an omnipotent deity placed the *Preliminary Discourse* firmly in the tradition of over a century of British natural philosophy. If secularization is a hallmark of modern science, in this sense Herschel's work has one foot planted in traditional natural philosophy: the study of nature Herschel advocates remains the study of *God's* nature. What the *Discourse* does, however, is make this religious aspect of natural philosophy more directly available to a wider audience. The scientific virtues of hope and humility are no longer the possession of the gentleman dilletante or aristocratic natural philosopher. Leisure or authority are no longer required in order to participate in the practice of natural philosophy; the power of scientific reasoning that gives rise to these spiritual insights is available to everyone. Pursuing natural philosophy brings about attitudes and habits of mind that provide "a most delightful retreat from the agitations and dissensions of the world, and from the conflict of passions, prejudices, and interests in which the man of business finds himself continually involved." An understanding of the methods of science cultivates virtues, and those virtues make science a retreat, whether from feuding in the Royal Society or the demands of a life of business—an ideal moral pursuit for the growing middle class.

Science, as Herschel defines it, "is the knowledge of many, orderly and methodically digested and arranged, so as to become attainable by one." Here, Herschel's use of the word *science* still holds an echo of its classical definition as a system of orderly knowledge (the science of theology, for instance, or the science of grammar). Throughout the *Discourse* though, Herschel applies the term almost exclusively to mean systems of knowledge about the physical world, and in particular a method of reasoning common to all such systems, using science in its modern connotation. This transition is on display in the titles of the three portions of the *Discourse* itself: though the book is *A Preliminary Discourse on the Study of Natural Philosophy*, the first section is on the nature and advantages to study of the *physical sciences*, the second is on the principles of *physical science*, and the third and final treats the divisions of *physics*. Beyond the title page, the content and even structure of the *Discourse* shows natural philosophy in the process of transitioning to modern science.[7]

Herschel builds his definition of science on the abstract, analytical reasoning that played such a significant role in his early career. If science is knowledge structured and arranged in order to be understood by the individual, this is only possible through a system of rational thought independent of the physical world. Logic and mathematics are such a system, which have the great utility of the conceptual clarity Herschel had been pursuing since his writings for the Analytical Society. Reasoning regarding mathematical concepts like "space, square, circle, a hundred" is clear, complete, and distinct. Using such terms introduces no misunderstandings or ambiguous shades of meaning into the process of reasoning, in contrast with words describing the physical world that can have different meanings depending on how the term is used and whether it is applied correctly. Terms with unclear meanings are a primary source of error in science, whereas mathematical reasoning allows one to mentally "walk uprightly and straightforward on firm ground" and gives a "proper and dignified carriage of mind" to scientific pursuits. The chain of mathematical reasoning lays out each link in the process of analysis, making it visible and accessible to all.[8]

Yet this dependence on mathematics for scientific reasoning introduced a tension into Herschel's conception of science. As his work with the Analytical Society showed, the ability to actually follow the mathematical reasoning required by certain fields of the physical sciences required "a degree of knowledge of mathematics and geometry altogether unattainable by the generality of mankind." Most of Herschel's readers had neither the time nor capacity for advanced mathematics, a pursuit difficult enough that only those "who devote to them their whole attention, and make them the serious business of their lives" could truly grasp the principles required. On the other hand, as Somerville had shown in explaining in conceptual terms the chain of argument of Laplace's great mathematical masterpiece and as Herschel had envisioned for his popular treatment of light, the "general train of reasoning" on which physical insights were based could still be explained to those without the requisite mathematical skills. The role of popularizers, then, becomes essential to Herschel's new vision of science: a reformed natural philosophy required both skilled mathematicians as well as those who could explain the principles and results of such to the general public.[9]

The ultimate goal of science, Herschel explained, is deriving laws of nature by applying mathematical reasoning to the physical world. Laws of nature, unlike human laws with their foundation on convention and historical contingency, are unchanging and unarbitrary. Herschel contrasted the work of the natural philosopher uncovering these laws to the situation of hypothetical "creatures of another world" (in modern parlance, aliens from another planet) set down in London with the task of deducing the laws of society. Such beings would face a nearly impossible task. As Herschel knew from his own abortive study of law, there were too many "incongruities, absurdities, and contradictions" in human laws to generalize into any overarching structure. This was not the case in nature, where the natural philosopher finds "no contradictions, no incongruities," no "maladministration" in nature's "sublime legislation." Herschel's days in the office of Mr. Saunders poring over legal cases provided his perspective for understanding the reasonable pursuit of nature's deeper, universal laws.[10]

Herschel then turned to the utility of the physical sciences in terms of human flourishing. Knowledge of nature allows for the unlocking of great physical power. This is most obvious with coal, the transformative technological effects of which were at this time becoming clear to all of Herschel's readers. Mathematical analysis could help readers grasp the scale of this new power. The ascent of Mont Blanc from Chamonix, for example, "considered, and with justice, as the most toilsome feat that a strong man can execute in two days," represents the same amount of work that could be accomplished with the combustion of just two pounds of coal. Yet no matter how much coal was mined, the physical power unlocked—as well as the distribution of goods this would allow—always remains limited because coal is itself consumed through usage. Scientific knowledge was quite different; it can travel further, be spread equitably, and is not, like coal, destroyed by use but rather honed and refined "passing through the minds of millions." Knowledge of the physical world, unlike physical wealth or power, is therefore radically egalitarian. A despot could monopolize material wealth or energy; they could not monopolize knowledge, which "can neither be adequately cultivated nor adequately enjoyed by a few." This meant that science should be disabused of everything keeping it exclusive or cloaked in obscurity, like the secrecy Herschel encountered with Fraunhofer and his methods of glass production. The difference, Herschel believed, between scientific knowledge and artisan or craft knowledge is that science must be open and accessible to anyone. Despite its terminology and "idioms of language," science must be stripped of everything tending toward obscurity, anything that led to "an appearance of superiority in its professors over the rest of mankind."[11]

The first portion of the *Discourse* was thus a call for a science open and accessible to all—and part of a reform movement that held moral as well as physical power. The practice of science would bring "into exercise a sufficient quantity of sober thought" that could then spread to social relations, legislature, and governance. When this happened, "the condition of the whole human species shall be permanently bettered." More than simply the application of science through technology to solve humanity's ills, this was instead Herschel's radically optimistic view of the reforming power of scientific thought. The virtues inculcated in those who practiced science could ultimately overcome the effects of human nature itself: "enabling the collective wisdom of mankind to bear down those obstacles which individual short-sightedness, selfishness, and passion, oppose to all improvements, and by which the highest hopes are continually blighted, and the fairest prospects marred." Science for Herschel was more than a tool to improve the human condition; thinking scientifically gave rise to virtues that could reform humanity itself.[12]

CREATURES OF REASON

With the claim that reasoning scientifically is the tool for overcoming even the brokenness of human nature, Herschel next turned to the rules of scientific reasoning, taking his readers into the processes of science to show how theories of nature are created. For most contemporary reviewers, this was the key portion of the book. Herschel himself called it the "cream of the work." Not since the writings of the English natural philosopher Francis Bacon had there been a systematic presentation of the methods of natural philosophy and the rules for building theories from observation. But whereas Bacon's goal was to make natural philosophy a branch of the state, vesting a small number of individuals with the power to create knowledge, Herschel's goal was the opposite: inviting all readers into the process of theory creation.[13]

Despite the importance of mathematical reasoning, Herschel acknowledged that experience was the only true source of knowledge of nature. Whereas experience in the form of passive observations might allow physical laws to be slowly pieced together from various occasions of witnessing processes at work in nature, rapid progress in science was only made through active observation—in other words, through experiment. Drawing readers into the laboratory through references to his own investigations, Herschel showed how natural philosophers used experiments to "cross-examine" nature. Like a lawyer, experiments allowed natural philosophers to interrogate their witness: "by comparing one part of his evidence with the other, while he is yet before us, and reasoning upon it in his presence," the experimenter put to nature itself "pointed and searching questions." Those branches of science opened to experiment, Herschel maintained, have had the most rapid development and given rise to the most complete theories.[14]

In the search for natural laws through experiment and observation, causes are what is truly sought. To understand any natural phenomenon, it must first be broken down as far as possible into its components, even though the cause of each of these may never be known. For example, the phenomenon of sound from a vibrating tuning fork can be reduced to a series of mechanical forces or impacts, even though the ultimate cause of mechanical force might remain unknowable. This potential inability to arrive at ultimate causes was not a problem for Herschel, who pointed out that everyone has direct knowledge of an intimately familiar yet unexplainable phenomenon: the exertion of force. We know we have the power to move our own limbs, and we experience such power, but we cannot explain where it comes from or how that power is communicated from will to movement. Mind is the agent and will is the cause, but the mechanism for this, the only direct act of causation we have immediate consciousness of, remains unknown. Ultimately, then, it is futile to inquire into causes themselves. Instead, the natural philosopher concentrates on laws that seem to be the immediate result of causes, looking for patterns in how the components of any phenomenon are related. It is only in this "modified and relative sense," never directly, that science can treat causes.[15]

Once the natural philosopher reaches the level of breaking phenomena into their constituent (but not ultimate) causes through experiment or observation, they can begin reasoning from these causes, and "the study of that phenomenon and of its laws becomes a separate branch of science." Causes then become themselves grounds for further reasoning and play "the same part in natural philosophy that axioms do in geometry." In other words, the natural philosopher can begin reasoning from *concepts* rather than phenomena. The goal of science then moves from the physical world to the mental: to arrive at laws that connect phenomenon and thus "transplant them out of the external world into the intellectual world, render them creatures of pure thought, and enable us to reason them out *a priori*." The natural philosopher can then reason *beyond* experience, can make predictions about a wider range of situations and combinations of causes than they could ever directly observe. The result is the *creation* of facts that do not necessarily correspond with any immediate experience. Herschel begins the *Discourse* reflecting on the physical weakness of humans in the natural world, yet despite this physical frailty, science is the means by which human *reason* can encompass the universe.[16]

There are no general rules for this process of determining scientific laws, but there are guides based on the experiences and successes of other natural philosophers. Herschel provided several methods for finding patterns in nature, many of which came from his own work on polarized light. For example, as Herschel knew from his work at Slough and the experiments he conducted with Arago in Paris, the "very vivid and beautiful colors" created by letting polarized light fall on certain crystals did not occur when the same light encountered fluids or opaque solids. Neither did it occur for *all* transparent crystals. By comparing the properties of crystals that did experience the effect, it was found to hold only for crystals that have the property of double refraction. Because there was no known exception to this, for Herschel this was an example of a generalization safe to treat as a physical law of nature. It could then be used as a means of reasoning toward a complete theory of light.[17]

Reasoning in this manner depends on careful and accurate classification. Organizing facts allowed the natural philosopher to see what different phenomena have in common. This was most effective when "facts are numerous, well observed, and methodically arranged," situations where the approach of gathering and comparing facts could employ, "as an engine, the division of labour." This division of labor distinguished mature sciences in which labor could be divided, such as astronomy, from less established branches of science that depended more on the work of skilled individuals. In the case of a science like astronomy, where techniques and systems already existed to organize observations, large numbers of observers could be trained to gather useful data. On the other hand, in sciences still developing, experts like Herschel were needed to set up and interpret experiments. A less developed science, such as optics, required "a union of many branches of knowledge in one person," whereas a more developed science could, as in industry, benefit from the division of labor. In such fields, a skilled scientific manager directed other trained observers, as Herschel did in his work on double and variable stars.[18]

Even when observations were outsourced to trained observers and their uniformity and precision ensured—for example, with the use of forms or worksheets created for recording observations—reasoning from those observations was still required to create scientific theories. Herschel provided a list of rules "guiding and facilitating" the search for laws of nature, which could then be used to construct theories. The first step of this process was looking for analogous examples among phenomena of the thing originally observed. If such an example is close enough to the original phenomenon, this similarity could become the basis for assuming a common cause. For example, if we know a force is required to keep a stone swinging in a sling in a circular path, then when the moon is observed circling Earth in a similar path, the natural philosopher can assume there must likewise be a force acting on the moon like the sling acts on the stone. The stone and the moon are analogous examples showing a common cause or law.[19]

The challenge is dealing with phenomena for which no clear analogies present themselves. In that case, the natural philosopher needed to search for other clues, of which Herschel offered several examples. These constitute Herschel's guide for inductive reasoning, and for many readers they were seen as the first modern attempt at providing systematic rules for scientific reasoning. For Herschel, however, they were simply practical guidelines. For instance, one of Herschel's rules is that if there is an invariable connection between phenomena (producing one always leads to the production of the other) or an invariable negation, then their causes are likely linked. Likewise, if increasing a certain effect leads to a proportional increase in the cause, the two are likely linked.

Following these guidelines or hints, the detection of a possible cause should lead to either a true cause that completely explains the facts or an abstract law showing that the two phenomena under investigation are invariably connected. These hints worked for reasoning with laws as well. Laws are of a higher order than any particular facts, but once determined they can themselves be grouped and classified in order to investigate how *they* are related. In other words, once new laws are established, they can be reasoned on just like phenomena and causes. Yet even as Herschel lays out guides to reasoning by induction to create scientific laws, he leaves the door open to hypothesizing more freely. His guides for reasoning are descriptive, based on his own experiences and those of his scientific colleagues, not proscriptive or meant to limit scientific reasoning. Herschel maintained that even though the rules he set out were how science usually progressed, the natural philosopher must not "be scrupulous as to *how* we reach to a knowledge of such general facts." Rather, they "must be content to seize them wherever they are to be found."[20]

Devising laws of nature is only half of the scientific process, however. The second half is verifying or testing those potential connections between phenomena and causes. In the case of a phenomenon that seems to confirm a well-known law, it is enough to examine the cases in which the cause is believed to be at work. For quantitative laws, this means measuring the law as precisely as possible in various instances. However, if reasoning leads to a new law, verification needs to be much more severe. Because laws should be universal in application, verification means extending the law to situations where it was not originally developed. The most powerful kind of verification comes when confirmations of a law spring up in unsought and surprising places, "in the course of investigations of a widely different nature from those which gave rise to the inductions themselves." For Herschel, epistemological surprise—the natural philosopher finding a law confirmed in an unexpected place or by a phenomenon that seems to have little connection with those from which the law was first formed—provides one of the most powerful confirmations that science is indeed uncovering true connections in the physical world.[21]

Herschel emphasized the strength of such law-like connections and the vision of nature it unfolds, inspiring readers with new perspectives offered by scientific reasoning and his view of a unified physical reality. The scope of the laws of nature extended from the realm of the minuscule to the immense, from the trade winds across the globe to the tiniest eddies in the smallest masses of air. In the sidereal universe, this romantic vista extended even farther. Astronomers can observe in the heavens, thanks to the work of Herschel and his father, star clusters "probably not less vast and complicated than our own" but reduced in size by distance until they seem crowded on each other. Simultaneously, within the atoms of a grain of sand, which might actually be as "remote from each other (proportionally to their sizes) as the stars of the firmament," Herschel speculated there could be "in that little microcosm, processes as complicated and wonderful as those of the great world around us." The romantic, all-encompassing view of science Herschel forged in the Alps and on the slopes of Etna became the goal of discovering nature's laws.[22]

The road to this grand conception of the universe is reasoning from observations of nature to laws of nature, but for Herschel the apex of science is an additional step, reasoning from those laws to the larger theories that unify and embrace them. Such unification, according to Herschel, had been reached in the physics of motion and force, where theories had moved from the realm of observation to the realm of pure reason. Theories, those conceptual frameworks for a particular field in which the laws of nature are encompassed, become "the creatures of reason rather than of sense." Their creation is the overarching goal of natural philosophy, the ultimate achievement through which human reason can truly encompass the universe. Because theories exist in the realm of rational thought, not the physical world, by forming them the mind moves in its own element, unhindered by constraints of observation or experiment. In creating theories, questions in science that remain unanswered, such as the nature of imponderables like light, heat, electricity, or gravity, can be reasoned on even if their physical causes remain invisible or undetectable. And though the creation of such theories might depend on people like Herschel who can apply complex mathematical analysis to laws arising from observations, understanding the process behind their creation is open to all. Comprehending a theory of nature gives the mind—even the mind of the general reader—more intimate contact with reality than facts communicated by senses. At the summit of the scientific endeavor, then, are theories that can be communicated to anyone. The result of natural philosophy certainly leads to both religious certitude and commercial benefit, but this is not its primary goal. The true achievement of science is a form of thought—thought that can arrive at true and intimate contact with reality.[23]

THE MAP OF KNOWLEDGE

Once he had explained "the spirits of the methods to which . . . natural science has been indebted" for its success, Herschel devoted the third and final portion of his work to exploring the different areas of science, showing how they are related, and illustrating his method at work within each. He begins with force, his example of an imminent effect with an unknown cause experienced by everyone, and shows how its analysis gave rise to mechanics, a field perfected by Newton on "sound induction" with rules that are general and complete regardless of any particular hypothesis on the ultimate structure of matter. The division of matter into solids, liquids, and gases leads to hydrostatics and pneumatics, this last an area in which the discovery of the weight of air was essential. Herschel highlighted the experiments of Blaise Pascal (1623–62), who had his brother-in-law Florin Périer (1605–72) carry a barometer to the top of Puy de Dôme, one of the extinct volcanoes Herschel had climbed in Auvergne, as a crucial step in pneumatics, "tending more powerfully than any thing which had previously been done in science to confirm . . . that disposition to experimental verification."[24]

Hydrostatics and pneumatics are based on mechanics applied to fluids, whereas analysis of solids gives rise to the realization that some solids have a disordered structure whereas others are ordered, which in turn gives rise to the science of crystallography. Like mechanics, the laws of crystallography are independent of assumptions regarding the nature of matter itself. Whatever the ultimate forces in crystals that arrange their constituent parts, it is clear those forces do not act the same way in every direction. This leads to the concept of *polarity*, seen on a large scale in magnetic needles. "It is not difficult," Herschel wrote, "if we give the reins to imagination, to conceive how attractive and repulsive atoms, bound together by some unknown tie, may form little machines or compound particles" with the property of polarity. But whether or not this atomic hypothesis is true, *polarity of matter* is one of the "ultimate phenomena" reached in the analysis of nature. In this way, Herschel was able to sidestep debates raging regarding light and the Laplacian program.[25]

The idea of motion communicated through bodies leads to discussions on the nature of sound and light. Herschel begins with a brief survey of acoustics, then moves to a more extensive discussion of light—both topics he had recently written treatises on for the *Encyclopedia Metropolitana*. Optical theory yields the telescope, which Herschel admits "must assuredly be ranked among the highest and most refined productions of human art," giving such an expanded sense of sight it might almost be considered a new sense entirely. Herschel outlines both the Newtonian theory of light as particles and the Huygensian theory of light as waves, an example of two contemporary theories offering different causes for phenomena. Polarized light provides a means of testing both of these theories. Indeed, for Herschel polarization is "an instrument of experimental enquiry . . . so marked and intimate" it might almost be said to be a new "kind of intellectual sense, by which we are enabled to scrutinize the internal arrangement of those wonderful structures which Nature builds . . . with a symmetry and beauty which we are never weary of admiring." What the telescope was to astronomy or the microscope to biology, polarized light has become to optics and crystallography. We feel, Herschel tells his readers, "we are on the eve of some extraordinary discovery, and expect every moment that some leading fact will turn up, which will throw light on all that appears obscure, and reduce into order all that seems anomalous."[26]

From light, Herschel moves to "cosmical phenomena," beginning with astronomy and celestial mechanics. Advancements since Newton have consisted of elaborating and accounting for all the details of gravity, which led to the important discovery of the stability of the solar system. Beyond the solar system, the immense distance of the stars "exalts them . . . into glorious bodies, similar to, and even far surpassing, our own sun, the centres perhaps of other planetary systems, or fulfilling purposes of which we can have no idea, from any analogy in what passes immediately around us." Herschel draws on his own work at the telescope to expand his vision of science. Stellar catalogs and surveys have "disclosed the existence of whole classes of celestial objects, of a nature so wonderful as to give room for unbounded speculation on the extent and construction of the universe." Stars have purpose, and if that is not to provide life and habitability to their own planets, then it is something else unimagined. Double stars, for instance, sweeping out immense orbits in the heavens, "must be accomplishing ends in creation which will remain for ever unknown to man."[27]

Herschel completes the unification of terrestrial with celestial phenomena by considering geology a "cosmical phenomenon." Geology, transformed into an inductive science during Herschel's lifetime, provides hints that Earth is an evolving system, a dynamic, living planet. Geologists no longer "bewilder their imaginations with wild theories" of early catastrophes that shaped the globe but now "aim at a careful and accurate examination of the records of its former state" in features of its surface and deposited strata. The key to this transition, which Herschel participated in during his travels in Italy and France, is that geologists no longer explain Earth's past with pure hypothesis but instead confine themselves to "careful consideration of causes evidently in action at present." The laboratory of geology, in which phenomena are linked to causes, is as large as Earth itself. Instead of bringing it to their workbench, careful observers must be sent out across the globe, transforming geology into the ideal imperial science. Information, Herschel noted, was already flowing in "respecting the geology of our Indian possessions, as well as of every other point where English intellect and research can penetrate."

This work, which needed to be coordinated on a large scale and depends on British imperial presence, should be propagated and supported "by the representatives of our national authority wherever our power extends." It was British military presence in Naples and Sicily, after all, that helped make Herschel's southern Italian expedition possible—and would soon make it possible for him to travel to the cape.[28]

Geology leads to mineralogy, a branch of science rich with philosophical speculation, connecting ideas on light with intermolecular forces. Here instrumentation was crucial for the field's development into a mature science, as Wollaston's goniometer, a tool for measuring angles in crystals, gave mineralogy solid quantitative footing. Along with chemical analysis, this meant mineralogy became a means of investigating the internal composition of matter. Chemistry is distinguished from mineralogy in that the former is the investigation of bodies with respect to their materials, whereas the latter investigates their structure. The dramatic changes produced by reactions makes chemistry ideal for displaying before audiences, Herschel admitted, and contributes to its popularity. Chemistry is perhaps "the most completely an experimental" science, with theories that are relatively simple and lead to "no profound mathematical researches." Nonetheless, atomic theory could be, after the laws of mechanics, the most important general theory science has yet disclosed. Chemical theory is also an example of scientific reasoning making things simpler as theory progresses, of "difficulties diminishing as we advance, instead of thickening around us in increasing complexity."[29]

Science, as the closing sections of the *Preliminary Discourse* made clear, was now global. Travelers can make important contributions through collections and observations, "but the resident alone can make continued series of regular observations" needed for careful inductive reasoning, such as observations of climate, tides, magnetic variations, and detailed geological notes. The promise of empire, in Herschel's mind, is that it will create observers across the entire world. "To what, then, may we not look forward, when a spirit of scientific enquiry shall have spread through those vast regions in which the process of civilization . . . is actually commenced and in active progress?" Scientific reason would not only inculcate morality and order into the minds of readers in Britain; science and its cultivation of rational thought would go hand-in-hand with colonial expansion. "And what may we not expect from the exertions of powerful minds called into action under circumstances totally different from any which have yet existed in the world, and over an extent of territory far surpassing that which has hitherto produced the whole harvest of human intellect?" British trade, British colonies,

British education, and British communication would create and then knit together observers in a global network feeding facts back to British theorists, just as British subjects labored to extract material wealth and direct it back to London. Despite Herschel's appeals for the international and nonpolitical nature of science, his vision of modern science was formed in an imperial context—one he would make use of when he decided to become a colonial resident himself in order to complete his own astronomical projects.[30]

As Herschel completed his scientific vision, science itself, like colonial expansion, remains apparently boundless, with many unexplored regions. There are only a few fields of science where laws allow full explanation of causes. Science remains where Newton saw himself, standing on the shore of a vast ocean. But this, Herschel believed, should only inspire more effort, an advance toward generality that is at the same time an advance toward simplicity. Nature only appears complicated, Herschel argued, when the investigator is "wandering and lost in the mazes of particulars" or working "downwards in the thorny paths of applications." Like the Alpine mountaineer, the methods of induction—of reasoning from phenomena to laws and onward to theories—provided a vantage point "from which we can take a commanding view" of nature. It is, Herschel concluded, a view in which "we never fail to recognize that sublime simplicity on which the mind rests satisfied that it has attained the truth."[31]

In almost all the areas of natural philosophy Herschel discussed— astronomy, geology, mineralogy, chemistry, electromagnetism, and optics—he wrote from personal experience as an experimenter and theorist. In his overall approach, especially in his belief that all the physical sciences are at work in a universe created by God, he was a natural philosopher. But in the particulars of each, in his dependence on mathematical analysis and standardized observations, in teasing out disciplinary boundaries, in generalizing the scientific method, and in his insistence that this practice is open to all, he is a scientist, though the word itself had not yet been coined. Though Herschel's reforming work bears the words "natural philosophy" in its title, in the process of setting out his vision Herschel was inventing modern science.

THE INVENTION OF SCIENCE

The *Preliminary Discourse* offered no new theories. It was not an *Origin of Species*. Yet it provided a model for theories to come, including Darwin's. For the first time, both general and genteel readers were provided a model of scientific practice and behavior—and this model was the same for all social classes. The reasons the *Discourse* was so effective in promulgating this vision was due not only to what was on the page: as much as the *Discourse* created a vision of science that would go on to influence Victorian society and the world, Herschel spent his career fashioning himself to be the person to write such a book. The author of the *Discourse* had made himself the ideal scientific practitioner, and his book thus became both apology and manual for pursuing a similar life.

Herschel's desire for reforms to British mathematics, nurtured since his student days at Cambridge, provided the foundation. That the human mind could grasp truth, that deductive certainty could be reached, and that reasoning on the physical sciences in their most developed form paralleled the axiomatic reasoning of mathematics—these certainties stemmed from Herschel's earliest days as a mathematical reformer. The tools of analysis—simplicity, uniformity, clarity of definitions—remained for Herschel the benchmarks of clear reasoning, the goal to which science strove. The ability of mathematics to explain the universe and provide predictions only later verified experimentally was for Herschel one of the strongest claims for his approach. Though many of his readers would only be able to appreciate this conceptually, Herschel's work in the Analytical Society meant he had pursued mathematical analysis deeply enough to see this for himself.

As a law student in London and through his correspondence with James Grahame, Herschel developed a view of law as a means of explaining the behavior of humanity. Laws were not universal arbitrators; they were based on experience, best-fit approximations for governing society. In some cases, especially in regard to capital offenses, they needed reform. The reality of human law both motivated Herschel's search for the deeper laws of nature and opened his eyes to the ways in which knowledge could help reform society. More than simply providing him a handful of analogies, his abortive study of law placed him firmly within the Whig tradition of civic virtue and reform and shaped the moral purpose the *Discourse* would ultimately fulfill.

The deductive, mathematical approach was only half of Herschel's vision of science though. The strength of his work was the way it married deduction with induction in the central, philosophical portions of the book, where Herschel provided a "theoretical" view of science, and in his descriptive narratives. Herschel's examples, data, observations, and experience gave flesh to the structure of the *Discourse*. The *Discourse* sat squarely on Herschel's early travels and self-fashioning. He wrote authoritatively about the physical sciences because he had explored them in his own laboratory and debated them with the leading natural philosophers of Europe. He used illustrations from geology not because he had read the relevant texts but because his ascent of Etna, his exploration of the extinct volcanoes of Auvergne, and his Alpine expeditions made him the authority consulted by those who went on to write those texts. He argued compellingly for the universalizing nature of science because he practiced that approach universally.

In addition to mathematical foundations and the breadth of his own experience, everything Herschel wrote about in the *Discourse* was built on a framework of precise observation and measurement. To bridge the connection between mathematical analysis and experience, observations had to be translated into data by precise and standardized measurement. Herschel learned this in his early years in London through his participation in the Astronomical Society and his own double star observations. His career in astronomy standardized and transformed the unique observations of his father into data accessible to other observers. The Astronomical Society's primary purpose was to unify astronomers in their approach to cataloging stars. Alongside businessmen and accountants like Baily and Babbage, Herschel poured over methods for reducing and recording astronomical data. Bringing a businesslike efficiency to the practice of science was a major difference between the Astronomical Society that Herschel had helped found and the traditional Royal Society, and it was the former approach Herschel championed in the *Discourse*.

Ultimately, the *Preliminary Discourse* served a significant moral function, in this respect making the transition from natural philosophy to modern science less an abrupt break than an important evolutionary step. Herschel viewed his apology for science as a means of cultivating personal and social virtue and combating atheism. Though Herschel's early views on organized religion were dubious, his time with the Stewarts moved him from a reserved broad church Anglicanism to a more personal faith. This was reflected in his arguments for natural philosophy as an inherently moral practice leading toward a deeper appreciation of the creator and away from skepticism and doubt, a theme of the *Discourse* over and above the virtuous effect he believed understanding the methods of science would have on society. The death of Margaret's sister Isabella in the course of writing sharpened this theme, leading Herschel to discuss the immortality of the soul, which his view of science made more, not less, rational.

Finally, the *Preliminary Discourse* was directed at scientific society itself, including some of his closest colleagues. Written against the backdrop of division and dissension, the *Discourse* was a missive aimed at moving science beyond the domain of aristocratic privilege. The controversies in the Royal Society of the 1820s and 1830s were about how natural philosophy would be conducted: whether it depended on positions of privilege for patronization or whether it was open (relatively speaking) to those with sufficient skill. On the one hand were Herschel, Babbage, and the members of the newer societies; on the other, Gilbert and his supporters. In a very real sense the *Discourse*, written from a position of retirement and withdrawal interrupted only to stand for election to the presidency against the Duke of Sussex, was a means of broadening the scientific process, making it public and accessible to those both inside and outside the walls of Somerset House. Natural philosophy, Herschel argued, could be pursued by anyone willing to carefully observe and learn what it took to transform those observations into data.

John Herschel's vision of natural philosophy as a tool for virtue in society and a means for a virtuous society was effective, and it was effective largely because he had fashioned himself as the ideal scientific practitioner. The irony is that even in its success, its vision of theorists surveying the different fields of physical science and generalizing from data produced all over the globe was over almost as soon as it had begun. Three years after the *Discourse* was published, the British Association for the Advancement of Science—a group arising from the reforming movement outside London and whose organizational structure mirrored the fields of physical science outlined in the *Discourse*—met at Cambridge. During discussions, Herschel's friend Whewell proposed a new term for those who practiced the sort of natural philosophy developing in the wake of the *Preliminary Discourse*: scientist. Within a generation, Herschel's personal polymathy was a thing of the past, and the first generation of professional scientists used the methods of the *Discourse* but divided science into fields in which his methods could be pursued independently. They wrested science from the dominance of the universities and the Anglican church and ultimately divorced it from its theological moorings. The first real triumph of Herschel's model of theory formation was Darwin's biological synthesis, modeled after the *Discourse*'s guidance—and this theory was ultimately problematic for Herschel. Though Herschel's work inspired the first generation of scientists, he remained the last of a generation of natural philosophers.[32]

The *Discourse* was the climax of Herschel's self-fashioning as the premier scientific practitioner of the nineteenth century, and it marked the conclusion of the first stage of a long life and career—encompassing his student days, his first experimental explorations in London, the pursuit of light, the beginnings of his astronomical career and the death of his father, European travels, engagement and marriage, and his rise to the apex of the London scientific community and subsequent withdrawal from it. With the publication of the *Discourse* and his defeat in the contest for presidency of the Royal Society, Herschel decided his vision of science would be vindicated not at the center of the scientific world but by pursuing his own projects at the very fringes of empire. The stage was thus set for Herschel's great scientific pilgrimage and his next endeavor, reforming the heavens themselves.

When Darwin and the *Beagle* arrived at harbor south of Cape Town, Herschel had been living for two years with his family in the country house of Feldhausen. His evenings were devoted to his astronomical sweeps, and his days were spent with Margaret collecting, painting, and categorizing botanical samples, ranging through the countryside, collaborating with astronomer Thomas Maclear (1794–1879) of the Cape Observatory, and consulting on educational arrangements for the colony. He was living his best life, far from the debates and debacles that had distracted him in England, able to devote himself exclusively to his scientific projects and his family. It was a lifestyle Darwin would emulate when he returned to England and started a family of his own. Herschel had his stars; Darwin would have his barnacles.

The young naturalist was unsure what to expect as he and FitzRoy traveled from Cape Town to call on the famous natural philosopher. Herschel's name had become scientific royalty; the king had knighted him Sir John prior to his departure for the Cape. Herschel was already a legend, whereas Darwin was an unknown naturalist. Despite his initial hesitation, the shy Darwin was immediately at ease in Herschel's presence, even a bit underwhelmed. He found his hero modest and reserved but easy to talk to and proud to show off the garden of native flowers he and Margaret had assembled, which Herschel had planted himself. "He never talked much," Darwin reflected on his visit, "but every word which he uttered was worth listening to."[33] Perhaps Herschel felt he had little left to say. He had spoken his piece in the *Discourse*, and in so doing endued natural philosophy with the contours of modern science.

EPILOGUE

After publication of the *Preliminary Discourse*, Herschel immersed himself in finishing what would become known as the Slough catalog, the complete revisiting, revising, and updating of his father's original nebular surveys. The work, which remained a unique endeavor in the astronomical community, took an immense amount of time and effort. Comprising the data from 428 sweeps made over a period of eleven years, the Slough catalog contained over 2,300 objects and represented the most complete catalog of non-stellar objects known. When it was complete, Airy, in awarding the work the Astronomical Society medal in 1833, said the catalog "established the beginning" of precision sidereal astronomy and represented "the first accurate account on which the knowledge of the yet growing bodies of the sky is to be founded."[1]

The Slough catalog, published like his father's original nebulae catalogs in the *Philosophical Transactions* of the Royal Society, represented the fruition of both William's and Caroline's astronomical legacies: the epitome of William's observational practice built on Caroline's meticulous organization. It was Caroline's reorganization of the entirety of William's nebulae observations from eight books of sweeps into a single listing that, Herschel admitted, made his resumption and continuation of his father's work even possible. He would, he said, never have undertaken it without Caroline's new master list, completed in her long retirement in Hannover. By publishing the Slough catalog, Herschel continued and perfected his family's tradition.[2]

But there was much more to be done. Even as Herschel's family grew with the birth of another daughter (Isabella, in 1831) and then a son (William, in 1833), Herschel remembered his glimpses of the southern skies from Sicily. Beyond the southern horizon waited an entire celestial hemisphere that had never been swept for nebulae, star clusters, or double stars. Herschel began seriously contemplating the extension of his father's work to the entire sky, relocating his family to a location south of the equator for the several years the project would require. When Herschel's mother died in 1832, he felt the time had finally come. He decided on the British colony of Cape Town, in South Africa, and as their departure date for this astronomical pilgrimage neared he raced to make corrections to the Slough catalog, proofing the final sheets for publication even as his telescopes were stowed for the voyage.[3]

Meanwhile, movements for reforming British science, culminating with the *Preliminary Discourse* but evolving without Herschel's direct involvement, continued. The defeat of the reforming party in the Royal Society helped prompt a movement among practitioners of science working outside of London and its environs: the creation of the British Association for the Advancement of Science. But Herschel was uninterested in this new project. When he was approached by the organizers of the association for his participation and support, he demurred. British natural philosophers were too independent, he had come to believe, to be coordinated into a cohesive body pursuing specific goals. And having narrowly escaped the loss of time and energy that would have resulted had he been elected Royal Society president, he believed even more strongly that the best service to science was devotion to an original line of research, not to coordinating scientific societies. He would not even agree to ask his colleagues in London what they thought about the plan for the new association. He was too busy.[4]

On Wednesday, 13 November 1833, Herschel, Margaret, their three children, nurse, servants, and John Stone, Herschel's mechanic, who would help him assemble and operate the twenty-foot telescope, embarked on the *Mount Stuart Elphinstone* for the Cape of Good Hope. Despite multiple offers from the government and the British Admiralty for transit on a naval vessel, Herschel resisted any official governmental support. He had resolved, he told his attempted supporters over and over again, that his expedition would be "the situation of an amateur embarking in a party of pleasure" or "an entirely irresponsible private adventure." Herschel and his family would not return to Britain again for more than four years, on 15 May 1838.[5]

Herschel's years at the Cape became the stuff of legend. His discoveries (as well as his purported discoveries, as in the case of the Great Moon Hoax) cemented his role as an exemplar of the vision of science he had advocated and created in the *Preliminary Discourse*. Besides his astronomical pursuits—which in addition to the completion of his celestial surveys included a careful study of the moons of Saturn, Halley's comet, sunspots, and variable stars—he continued the universalizing, physicalizing approach he had begun in the Alps and among the volcanoes of Italy: measuring tides, geologizing, and gathering meteorological data. With Margaret, he also began a project of documenting Cape flora, in addition to participating in the education of his growing family. He would look back on his time at the Cape as among the happiest years of his life.[6]

Herschel's Cape years were also a kind of scientific apotheosis that only increased his standing in the world of natural philosophy: when the young Charles Darwin called on him with the captain of the *Beagle*, it was very much as a pilgrim devotee. Yet the Cape was an education for Herschel as well. There he saw firsthand the effects of colonization and the continuing Xhosa Wars between European settlers and the original inhabitants of the Cape region. He met colonial governors on their way to posts in India or to positions at the Cape and conversed and corresponded on policy, missionary work, and education—and on occasion refusing to sit at the same table with officials who he felt had stolen land from indigenous peoples. When he returned to England, Herschel carried with him an altered view of science as an imperial tool and a new perspective of the effects of colonialism at the frontiers of empire.[7]

While Herschel was away from London, the reforming initiatives in the Royal Society did not cease. Rather, and to the surprise of many, the Duke of Sussex began to implement many of the reforms Herschel and his colleagues had urged. Aided by the longtime society treasurer, John William Lubbock (1803–1865), it seemed the duke was in fact the sort of institutional reformer Herschel had ultimately hoped for—reforming the society from the inside and placing it "upon a progressive and active footing." In 1846 the number of fellows admitted annually was finally capped. These changes did not make Herschel any more interested in the role of president, however. In 1848, when the presidential chair became vacant, Herschel was once again the favored choice. Though his election this time would likely have been assured, he immediately responded that his residency and occupations made it impossible for him to take on the responsibilities. (He was then living in Kent, even farther from London than Slough, and unwilling to give up his domestic retreat.) For him, science was still best served in his personal pursuit.[8]

Despite his continued aloofness from scientific society, Herschel's career after the Cape flourished. He rose to the minor nobility, being made a baronet by Queen Victoria at her coronation in June of 1838, soon after his return from the Cape. He continued as Britain's unofficial spokesman of science and ultimately became a leading figure in the

British Association—presiding over its 1845 meeting at Cambridge—despite his early resistance. In 1847 he completed the calculations and reductions required for his massive *Results of Astronomical Observations Made during the Years 1834, 5, 6, 7, 8, at the Cape of Good Hope; Being the Completion of a Telescopic Survey of the Whole Surface of the Visible Heavens, Commenced in 1825* (almost always abbreviated as the *Cape Results*), the culmination and completion of the Herschellian astronomical program. Symbolically, he moved into the shadow of Isaac Newton, taking a position late in his career, like Newton, as master of the mint, where the work he did to reform and modernize that ancient institution resulted in serious mental and physical health issues. (The parallels with Newton would continue to his death, as Herschel was buried close to Newton's tomb in Westminster Abby.)

Even as his other writings grew in influence and were translated abroad, the *Preliminary Discourse* remained the articulation of science for the age, and Herschel continued as its embodiment. The story of his time at the Cape and the remainder of his long life and career in England afterward remains to be told. But it followed a trajectory established, defined, and outlined by his invention of science, the completion of a personal vision created during his early career and spelled out in his *Preliminary Discourse*.

NOTES

INTRODUCTION

1. John Herschel, "A Review of *The Principles of Fluxions*, by William Dealtry," Herschel Family Papers, Series I, Subseries A, Works, 1811–1871, Harry Ransom Center, University of Texas, Austin (henceforth HRC), container 8.21.

2. Herschel, *Preliminary Discourse*, 190.

3. Cunningham and Williams, "De-centring the 'Big Picture'"; and Cunningham, "Getting the Game Right." For the *Discourse* and the philosophy of science see, for instance, Cobb, "Is John F. W. Herschel an Inductivist?"; Cobb, "Inductivism in Practice"; and Good, "John Herschel's Optical Researches." Herschel's *Discourse* has also been the subject of several dissertations, including Lal Jain, "Methodology and Epistemology"; Panter, "Sir John F. W. Herschel and Scientific Thought"; Bolt, "John Herschel's Natural Philosophy"; and most recently, Lukas M. Verburgt, "Herschel's Philosophy of Science," in Case and Verburgt, *Cambridge Companion to John Herschel*. For Herschel's influence on Whewell, see Yeo, *Defining Science*.

4. Secord, "The Conduct of Everyday Life: John Herschel's *Preliminary Discourse on the Study of Natural Philosophy*," in Secord, *Visions of Science*, 80–106. For Darwin's influence on Herschel see Ruse, "Darwin's Debt to Philosophy"; Schweber, "John Herschel and Charles Darwin"; and most recently Pence, "Sir John F. W. Herschel and Charles Darwin."

5. Lubenow, *"Only Connect,"* 273.

6. On this debate, see Cunningham and Williams, "De-centring the 'Big Picture'"; Cunningham, "Getting the Game Right"; Schaffer, "Scientific Discoveries"; and Harrison, *Territories of Science and Religion*.

7. Previous biographical studies include Buttmann, *Shadow of the Telescope*; Shorland, *Sir John F. W. Herschel*; and Snyder, *Philosophical Breakfast Club*.

PROLOGUE: THE ASTRONOMER AND THE AUTOCRAT

1. Crosland, *Science under Control*, 23, 35; Hahn, *Pierre Simon Laplace*, 84.

2. Herschel, *Scientific Papers*, lxi; Crosland, *Society of Arcueil*, 159; Rudwick, *Bursting the Limits of Time*, 46; Gleason, *Royal Society of London*, 3.

3. Crosland, *Society of Arcueil*, 420.

4. Hahn, *Pierre Simon Laplace*, 118; Secord, *Victorian Sensation*, 56.

5. Rudwick, *Bursting the Limits of Time*, 379, 388.

6. Rudwick, *Bursting the Limits of Time*, 390, 465.

7. Herschel, *Scientific Papers*, lxi–lxii.

8. Hoskin, *Discoverers of the Universe*, 180; John Herschel to Johann Wilhelm A. Pfaff, 5 July 1824, Royal Society Herschel Papers (henceforth RS:HS) 20.182.

CHAPTER 1: A CAMBRIDGE DISCONTENT

Epigraphs: Quoted in Dubbey, "Introduction of the Differential Notation," 38; quoted in Ball, *History of Mathematics at Cambridge*, 113.

1. The outlines of this conflict are given in Becher, "Radicals, Whigs, and Conservatives." Becher quotes from the letters of Thomas Greenwood, in whose rooms at Cambridge this exchange purportedly took place.

2. Warwick, *Masters of Theory*, 53–56.

3. Crowe, *Extraterrestrial Life Debate*, 63–67.

4. Quoted in Heilbron, "Mathematician's Mutiny," 91. On Herschel being fit for Bedlam, see Lubbock, *Herschel Chronicle*, 99, 103–4, and 179.

5. Hoskin, *Construction of the Heavens*, 62–63.

6. For background on the Baldwin family, see Hoskin, "Mary Herschel's Fortune." On the life and work of Caroline Herschel, see Winterburn, *Quiet Revolution*.

7. Buttmann, *Shadow of the Telescope*, 8–9. See also Emily Winterburn, "John Herschel: A Biographical Sketch," in Case and Verburgt, *Cambridge Companion to John Herschel*.

8. For an excellent and accessible overview of the conceptual difference between Newton's approach to calculus and the Continental approach originated by Gottfried Wilhelm Liebniz (1646–1716), see Richards, *Generations of Reason*, 162–65.

9. Colley, *Britons*, 286.

10. Crosland and Smith, "Transmission of Physics," 7.

11. Crosland and Smith, "Transmission of Physics," quoted on 11.

12. For William's trips, see Winterburn, "The Herschels," 95. On hiring Rogers, see Hoskin, "Mary Herschel's Fortune," 221.

13. Warwick, *Masters of Theory*, 53; Linehan, *St John's College*, 230.

14. Desmond and Moore, *Darwin's Sacred Cause*, 9.

15. On Herschel's student years, see Shorland, *Sir John F. W. Herschel*, 16–17. The value calculator at MeasuringWorth.com estimates the purchasing power of £60 in 1809 to be approximately £4,300 today. In 1792 a coach service between London and Cambridge had been initiated. On travel time, see Linehan, *St John's College*, 217.

16. Warwick, *Masters of Theory*, 53–56, 143.

17. "Woodhouse," 500; Richards, "Rigor and Clarity," 310; quote from Woodhouse, *Treatise on Plane and Spherical Trigonometry*, iv.

18. Dubbey, "Introduction of Differential Notation," 40. See also Topham, "Science, Print, and Crossing Borders." To calculate current value, see "Measuring Worth," accessed 21 September 2021, https://www.measuring worth.com/calculators/ukcompare/. For details on Babbage's purchase, including discussion of the date of purchase, which Babbage recounts differently in different contexts, see Fisch, "Babbage's Two Lives," 103–4.

19. Becher, "Whewell and Cambridge Mathematics," 3.

20. Alexander Rogers to John Herschel, 5 March 1810, RS:HS 14.394.

21. Alexander Rogers to John Herschel, 6 January 1812, RS:HS 14.396.

22. Linehan, *St John's College*, 219; James Grahame to John Herschel, 19 or 26 January 1812, RS:HS 8.240.

23. Watts, "'We Want No Authors,'" 410–11.

24. Herschel, "Analytical Formulae," 133; Herschel, "Trigonometrical Formulae."

25. Boyd, *Mad, Bad, Dangerous People*, 181. For a contemporary account of the furor around the formation of the Bible Society, see Gunning, *Reminiscences of the University*, 279–80.

26. Enros, "Analytical Society," 26–28. For details on Bromhead, see Edwards, "Bromhead."

27. James Grahame to John Herschel, June 1812, RS:HS 8.244; John Herschel to Charles Babbage, 1 July 1812, RS:HS 2.2.

28. Herschel, "On a Remarkable Application of Cotes's Theorem," 26.

29. Herschel, "On a Remarkable Application of Cotes's Theorem," 10n.

30. John Herschel travel diary, Herschel Family Papers, Series I, Subseries B, Personal Records, 1801–1871, HRC container 22.1.

31. John Herschel travel diary, HRC container 22.1. For the friendship between Watt and William Herschel, see Lubbock, *Herschel Chronicle*, 234.

32. John Herschel travel diary, July 1810, HRC container 22.1. For biographical information on Robert Grahame, see Boase, *Modern English Biography*, 1199.

33. Shorland, *Sir John F. W. Herschel*, 18; John Herschel travel diary, 9 August 1810, HRC container 22.1.

34. John Herschel travel diary, August 1810, HRC container 22.1. The most extensive account of Grahame's life and career can be found in the memoir at the beginning of Grahame, *History of the United States*.

35. Evans, *Forging of the Modern State*, 61; James Grahame to John Herschel, 16 January 1812, RS:HS 8.239.

36. Warwick, *Masters of Theory*, 65.

37. James Grahame to John Herschel, 28 January 1813, RS:HS 8.248; Alexander Rogers to John Herschel, 14 June 1813, RS:HS 14.398.

CHAPTER 2: A BRANCH OF ELEGANT LITERATURE

Epigraph: Quoted in Enros, "Analytical Society," 41.

1. For details on these events see Gillen, *Assassination of the Prime Minister*; and Linklater, *Why Spencer Perceval Had to Die*.

2. John Herschel to John William Whittaker, 7 July 1814, St John's College, Cambridge; John Herschel to John William Whittaker, 20 February 1813, St John's College, Cambridge; John Herschel to Charles Babbage, 24 October 1814, RS:HS 2.31.

3. John Herschel to Charles Babbage, 27 June 1813, RS:HS 2.14/20.9; Lubenow, *"Only Connect,"* 274.

4. John Herschel to Charles Babbage, 8 February 1813, RS:HS 2.6/20.5; Charles Babbage to John Herschel, 13 June 1813, RS:HS 2.13. The published paper was Herschel, "On Trigonometrical Series," 33–64.

5. Herschel, "On Equations of Differences," 65–114.

6. John Herschel to Charles Babbage, 25 February 1813, RS:HS 2.8/20.6.

7. John Herschel to Charles Babbage, 27 June 1813, HS 2.14/20.9.

8. "Preface," *Memoirs of the Analytical Society*, i–ii.

9. "Preface," *Memoirs of the Analytical Society*, xxi.

10. This early draft is manuscript James.512, St John's College, Cambridge. "Preface," *Memoirs of the Analytical Society*, xvi.

11. Herschel, "On Trigonometric Series," 34; Herschel, "On Equations of Differences," 91, 86.

12. "Preface," *Memoirs of the Analytical Society*, iv, xxi.

13. John Herschel to Charles Babbage, 25 July 1813, RS:HS 2.16/20.10. Instructions to the printers provided in the drafts of Herschel's papers in manuscript James.512, St John's College, Cambridge.

14. Manuscript James.512, St John's College, Cambridge. For details of the delay in printing, see Charles Babbage to John Herschel, 1 May 1813, RS:HS 2.10; John Herschel to Charles Babbage, 4 May 1813, RS:HS 2.11/20.8; and Charles Babbage to John Herschel, 25 May 1813, RS:HS 2.12.

15. John Herschel to Charles Babbage, 13 October 1813, RS:HS 2.19/20.15.

16. John Herschel to Edward Bromhead, 19 November 1813, quoted in

Enros, "Analytical Society," 41. On lack of reviews, see Enros, "Analytical Society," 37.

17. John Herschel to Charles Babbage, 13 January 1814, RS:HS 2.20/20.15.

18. John Herschel to Charles Babbage, 13 January 1814, RS:HS 2.20/20.15.

19. Alexander Rogers to John Herschel, 20 February 1814, RS:HS 14.399.

20. John Herschel to John Whittaker, 2 July 1813, St John's College, Cambridge.

21. James Grahame to John Herschel, 9 March 1813, RS:HS 8.249; Bayly, *Birth of the Modern World*, 145; Boyd, *Mad, Bad, and Dangerous*, 144.

22. Mary Pitt Herschel to John Herschel, 10 November 1813, Herschel Family Papers, Series I, Subseries C, Correspondence, 1811–1871, HRC H/M-0619. William set out his argument in a letter to John two days earlier. See William Herschel to John Herschel, 8 November, 1813, published in Lubbock, *Herschel Chronicle*, 348–50.

23. Herschel, *Preliminary Discourse*, 77.

24. James Grahame to John Herschel, 5 November 1813, RS:HS 8.258, and 10 January 1814, RS:HS 8.260.

25. Boyd, *Mad, Bad, and Dangerous*, 13, 152.

26. John Herschel to William Herschel, February 1814, HRC L-0546.1.

27. James Grahame to John Herschel, 10 January 1814, RS:HS 8.260.

28. Golinski, *Science as Public Culture*, 236, 269.

29. "Clarke, Edward Daniel."

30. John Herschel to William Herschel, February 1814, HRC L-0546.1; and John Herschel to Charles Babbage, 26 January 1814, HRC H/L-0047. On Herschel's purchases see Snyder, *Philosophical Breakfast Club*, 44.

31. Charles Babbage to John Herschel, 4 July 1814, RS:HS 2.24; John Herschel experimental notebook, vol. 1, 26 April 1814, Science Museum Group Collection (hereafter SMGC), MS/0478/1. On nomenclature, see, for example, John Herschel experimental notebook, vol. 1, 6 March 1814.

32. Herschel, "Consideration of Various Points of Analysis," 440.

33. Golinski, *Science as Public Culture*, quoted on 266.

34. John Herschel to Charles Babbage, 20 September 1814; John Herschel experimental notebook, vol. 1, 5 September 1814 and 1 November 1814, SMGC, MS/0478/1; John Herschel to Charles Babbage, 23 March 1815, HRC H/L-0050; John Herschel to Charles Babbage, 20 September 1814, RS:HS 2.30/20.19; James Grahame to John Herschel, 19 July 1814, RS:HS 8.263 and 8 March 1815, RS:HS 8.267.

35. On Smithson, see "Tennant, Smithson."

36. John Herschel to Charles Babbage, 23 March 1815, HRC H/L-0050. For details on Cumming, see "Cumming, James."

37. John Herschel to Charles Babbage, November 1814, RS:HS 2.33/20.21.

38. For a discussion of perceptions of mental health during this period,

especially in relationship to natural philosophers, see Musselman, *Nervous Conditions*. John Herschel to Charles Babbage, 10 September 1815, RS:HS 2.39/20.24 and 24 September 1815, RS:HS 2.40/20.26. Notes from his stay include more than forty pages of a draft on a paper on functional equations (HRC container 8.8). This paper was Herschel, "On the Development of Exponential Functions"; drafts are manuscripts James.516 and 512, St John's College, Cambridge. For Herschel's attempts to transform thought, see Ashworth, "Memory, Efficiency, and Symbolic Analysis."

39. John Herschel to Charles Babbage, 7 October 1815, RS:HS 2.42/20.27 and 6 November 1815, RS:HS 2.44/20.28. Grahame, still practicing in Scotland, hoped Herschel would reconsider his decision, telling his friend the study of law would provide an "anchor" and "resting place" from which to pursue his other studies (James Grahame to John Herschel, 3 November 1815, RS:HS 8.269).

40. Charles Babbage to John Herschel, 13 November 1815, RS:HS 2.47; John Herschel to Charles Babbage, 18 December 1815, RS:HS 2.51/20.30. The single exam question Herschel did contribute to that year's exam is an early signal of his growing interest in the area he would eventually see as the intersection between analysis and its physical application: optics (in particular, his question asked for a mathematical explanation of double refraction).

41. Becher, "Radicals, Whigs, and Conservatives," quote on 422, 416; Craik, "Calculus and Analysis in Early 19th-Century Britain," quote on 249; Playfair, Review of *Traité de mécanique céleste*, 281.

42. John Herschel to Charles Babbage, 6 January 1816, RS:HS 2.54/20.31.

43. John Herschel to Charles Babbage, 7 February 1816, RS:HS 2.55/20.32.

44. Fisch, *Creatively Undecided*, 151. For more details of this conceptual transformation, see also Fisch, "'Emergency Which Has Arrived,'" 267; Richards, "Rigor and Clarity," 310; and Koppleman, "Calculus of Operations," 180.

45. John Herschel to Charles Babbage, 14 July 1816, RS:HS 2.64/20.35.

46. Lacroix, *Elementary Treatise*. Herschel's *Collection of Examples* would be published in 1820. George Peacock to John Herschel, 3 December 1816, RS:HS 13.247.

47. George Peacock to John Herschel, 4 March 1817, RS:HS 13.248. For the perceived dangers of analysis to Anglican conservatives at Cambridge, see Ashworth, *Trinity Circle*, esp. 18–34.

48. William Whewell to John Herschel, 6 March 1817, RS:HS 18.158.

49. For details on William's plans near the end of his life, see Hoskin, "William Herschel's Agenda."

50. On the geology of Devonshire, see Bord, *Science and Whig Manners*, 36.

51. John Herschel to John William Whittaker, 2 September 1816, St John's College, Cambridge.

52. John Herschel experimental notebook, vol. 1, 9 December 1816, SMGC, MS/0478/1. Royal Astronomical Society, Herschel Archive, MSS

Herschel J. 1.1/1, journal no. 1. Hoskin notes that Herschel began work aligning the twenty-foot telescope on the meridian as early as September 17 of that year and that the instrument had not been used by William since August 1814. Hoskin, "William Herschel's Agenda," 448.

53. John Herschel to Charles Babbage, 10 October 1816, RS:HS 2.68.

54. Craik, "Polylogarithms, Functional Equations, and More." Spence's 1809 work was Spence, *Essay on the Theory of the Various Orders of Logarithmic Transcendents*. Spence, *Mathematical Essays*.

55. John Herschel to Charles Babbage, 30 January 1817, RS:HS 2.71.

56. George Peacock to John Herschel, 17 March 1817 and 30 May 1817, RS:HS 13.249 and 13.250. Herschel's father wrote him a letter detailing how to go about having the mirrors cast. William Herschel to John Herschel, 5 May 1817, Royal Astronomical Society, Herschel Archive, MSS Herschel W. 1/11.4; John Herschel to Charles Babbage, n.d., RS:HS 2.77.

57. George Peacock to John Herschel, March 1818, RS:HS 13.258. What remains of Herschel's *Algebra* can be found on the reverse of sheets of mathematical calculations held in Herschel Family Papers, HRC container 8.2.

58. John Herschel to Charles Babbage, 7 August 1814, RS:HS 2.28/20.18; Herschel, "Mathematics," in Schweber, *Aspects of the Life and Thought*, 436.

CHAPTER 3: REVOLT OF THE ASTRONOMERS

Epigraph: Schweber, *Aspects of the Life and Thought*, quoted on 438.

1. Crosland, *Society of Arcueil*, 1, 420.

2. For extracts from these letters see Lubbock, *Herschel Chronicle*, 316–18.

3. On the earlier proposed trip, see Charles Babbage to John Herschel, 20 July 1816, RS:HS 2.65. Banks quoted in Gascoigne, *Science in the Service of Empire*, 155.

4. Herschel, "On the Hyposulphurous Acid." For details on Herschel's chemistry-related work, see "Herschel and Hypo," in Ross, *Nineteenth-Century Attitudes*, 173–93; and Kelley Wilder, "Photology, Photography, and Actinochemistry: The Photographic Work of John Herschel," in Case and Verburgt, *Cambridge Companion to John Herschel*.

5. Herschel, "On the Application of a New Mode of Analysis," 23–33, 24.

6. Herschel travel journal, HRC container 22.2; John Herschel to Mary Pitt Herschel, 17 Jan 1819, HRC H/L-0514.3.

7. Hahn, *Pierre Simon Laplace*, 165–66.

8. John Herschel experimental notebook, vol. 1, 6 October 1814, 29 November 1814, and March through June 1818, SMGC, MS/0478/1. William Whewell to John Herschel, 19 June 1818, RS:HS 18.159. Herschel had been interested in double refraction since at least 1815, when he had proposed an exam question on the topic to his pupils at Cambridge. John Herschel to Charles Babbage, 18 December 1815, RS:HS 2.51.

9. Crosland and Smith, "Transmission of Physics," 8.

10. On the Académie, see Crosland, *Science under Control*; and Fox, *Savant and the State*. On hearing papers read at the meetings, Crosland, *Society of Arcueil*, quoted on 155. The quote continues, "For what benefit can be derived from the hearing of mathematical calculations . . . the comprehension of which can only be the result of patient study?"

11. For William's election to the Académie, see Crosland, *Science under Control*, 385.

12. Herschel listed everyone he met at this meeting in his notes regarding this trip in Herschel travel journal, HRC container 22.2.

13. Crosland, *Society of Arcueil*, 221.

14. John Herschel to Charles Babbage, 19 February 1819 and 25 March 1819, RS:HS 2.110/20.66 and 2.113/20.68. He described his experimental apparatus in John Herschel, "On Certain Optical Phenomena Exhibited by Mother-of-Pearl."

15. Charles Babbage to John Herschel, 19 February 1819, RS:HS 2.111.

16. John Herschel to William Whewell, 19 August 1818, RS:HS 20.56; John Herschel to David Brewster, 24 November 1819, RS:HS 20.76; John Herschel to Charles Babbage, n.d., RS:HS 2.121/20.73, emphasis added. For details of Herschel's optical work during this period, see Good, "John Herschel's Optical Researches."

17. Herschel, *Preliminary Discourse*, 354.

18. Herschel, "On the Action of Crystallized Bodies," 47 and 64.

19. Herschel, "On the Action of Crystallized Bodies." Herschel did believe such research could help explain the internal structure of crystals. See, for instance, Herschel, "On Certain Remarkable Instances," 21.

20. Schweber, *Aspects of the Life and Thought*, 434–35. For more on the extent of Herschel's pre-1819 formalism, see Ashworth, "Memory, Efficiency, and Symbolic Analysis."

21. For Baily, see Florence et al., "Baily, Francis"; De Morgan, "Account of a Correspondence"; and Richards, *Generations of Reason*, 230–38.

22. Heilbron, "Mathematicians' Mutiny."

23. Baily, appendix to Cagnoli, *Memoirs on a New and Certain Method*, 30–31.

24. Baily, appendix to Cagnoli, *Memoir on a New and Certain Method*, 35.

25. Dreyer and Turner, *History of the Royal Astronomical Society*, 21.

26. Herschel diary, 12 January 1820, HRC container 16.7; Ashworth, "Calculating Eye."

27. John Herschel to Charles Babbage, 17 January 1820, RS:HS 2.125/20.81; Herschel diary, 17 January 1820, HRC container 16.7.

28. Herschel, "Address of the Society," 4–5.

29. Herschel, "Address of the Society," 5.

30. Herschel, "Address of the Society," 6. For British instrument makers, see Taylor, *Mathematical Practitioners*, 43.

31. Herschel, "Address of the Society," 7; Francis Baily to John Herschel, 11 March 1820, RS:HS 3.37.

32. Francis Baily to John Herschel, 11 March 1820, RS:HS 3.37.

33. John Herschel to Charles Babbage, 19 March 1820 and 20 April 1820, RS:HS 2.130/20.85 and 2.134/20.91.

34. Herschel diary, 13 March 1820, HRC container 16.7; John Herschel to Charles Babbage, 12 August 1820, RS:HS 2.142/20.99.

35. Charles Babbage to John Herschel, 22 May 1820, RS:HS 2.136; John Herschel to Charles Babbage, 12 July 1820, 19 December 1820, and 22 January 1821, RS:HS 2.138, 2.150/20.106, and 2.153/20.110.

36. For details on these instruments and results, see Herschel and South, "Observations of the Apparent Distances."

37. Herschel, "On the Aberrations of Compound Lenses," 224.

38. Quoted in Gilbert, "Election to the Presidency," 259.

39. Herschel diary, 20 and 21 June 1820, HRC container 16.7.

40. "Report of the Council," 32.

CHAPTER 4: VALLEYS AND SUMMITS

1. John Herschel to Charles Babbage, 7 August 1814, RS:HS 2.28.

2. Herrmann, "Drawings by Sir Joshua Reynolds," 650. The purchase cost was 183 pounds, 14 shillings, 6 pence. Value conversion calculated with https://www.measuringworth.com/calculators/ukcompare/, which gives the value of this amount as approximately 16,000 pounds in 2020. According to Herrmann, this purchase remains the largest collection of the Reynolds's drawings in existence.

3. Mary Pitt Herschel to John Herschel, 6 May 1821, HRC H/M 620.2.

4. Herrmann provides evidence that two of the three sketchbooks Herschel bought were purchased on behalf of "Miss Gwatkin." The portrait of Reynolds was sold by John Herschel's grandson, the Reverend Sir John Herschel, in the late 1930s, and its present whereabouts are unknown. Herrmann, "Drawings by Sir Joshua Reynolds," 650n8, and 654.

5. John Herschel to Charles Babbage, 2 July 1821, RS:HS 20.120.

6. Charles Babbage to John Herschel, 20 May 1821, RS:HS 2.160.

7. John Herschel to Charles Babbage, 2 July 1821, RS:HS 20.120; Charles Babbage to John Herschel, 5 July 1821, RS:HS 2.163.

8. Mary Pitt Herschel to Charles Babbage, 9 July 1821, RS:HS 20.121.

9. Mary Pitt Herschel to John Herschel, 10 July 1821, HRC H/M 620.3. Regarding Herschel's inheritance from his Aunt Mary, see Hoskin, "Mary Herschel's Fortune," 215. Charles Babbage to Mary Pitt Herschel, n.d., transcription of letter appears between RS:HS 20.119 and 20.120. In Mary's letter of 10 July, she tells Herschel she is "happy to hear that you have put your in-

structions with regard to your marriage settlement into Mr. Bladgrove's hands" (HRC H/M 620.3).

10. Edgeworth, *Letters from England*, 597; Herrmann, "Drawings by Sir Joshua Reynolds," 653. Whatever the cause of the break, it must have been reconcilable, as Edgeworth records Theophila Gwatkin and one of her daughters as guests at a dinner Herschel and his wife hosted in March 1831. Edgeworth, *Letters from England*, 500–503. On the mention of Herschel's purchase in the papers, see Charles Babbage to John Herschel, 21 May 1821, RS:HS 2.161.

11. Rudwick, *Bursting the Limits of Time*, 431, 444; John Herschel to Charles Babbage, 12 July 1820, RS:HS 2.138; Good, "John Herschel's Travels," 286.

12. Rudwick, *Bursting the Limits of Time*, 431, 497, 537.

13. John Herschel to William Herschel, 23 July 1821, HRC H/L-0551.

14. John Herschel to Henry Kater, 5 August 1821, RS:HS 20.123; Herschel travel journal, 28 July 1821, HRC container 22.3.

15. Hahn, *Pierre Simon Laplace*, 165–66.

16. Herschel records Laplace's quote as "Il me semble que les idees de M votre Pere suis trues philosophiques & tres vrais" (Herschel travel journal, 29 July 1821, HRC container 22.3).

17. Dettelbach, "Humboldtian Science," 289.

18. Good, "A Shift of View," 42. See also Gregory A. Good, "Herschel's Planet: Earth in Cosmic Perspective," in Case and Verburgt, *Cambridge Companion to John Herschel*.

19. Schaaf, *Tracings of Light*. More than three hundred of Herschel's drawings remain in the Graham Nash Collection, which were donated to the Getty Museum in 1991. See also Omar W. Nasim, "Drawing Observations Together: John Herschel and the Act of Drawing in Scientific Observations," in Case and Verburgt, *Cambridge Companion to John Herschel*.

20. John Herschel to William Herschel, 12 August 1821, HRC H/L-0552.

21. John Herschel to James A. Gordon, 15 August 1821, RS:HS 20.124; John Herschel to William Herschel, 12 August 1821, HRC H/L-0552.

22. Rudwick, *Bursting the Limits of Time*, 14, 42, 72; Rudwick, *Worlds before Adam*, 105.

23. See, for instance, Hevley, "Heroic Science of Glacier Motion"; Reidy, "John Tyndall's Vertical Physics"; and Kaalund, "A Frosty Disagreement."

24. Herschel's measurements from the trip can be found in Herschel, "*Barometrical* and Thermometrical Observations, *Itinerant . . .* ," Series I, subseries A, Works, 1811–1871, HRC container 9.5. On his night wanderings, see Herschel travel journal, 21 July–31 August 1821, HRC container 22.3.

25. John Herschel to William Herschel, 30 August 1821, HRC H/L-0553.

26. John Herschel to William Herschel, 30 August 1821, HRC H/L-0553; John Herschel to Mary Pitt Herschel, 15 September 1821, HRC H/L-0515.4.

27. John Herschel to William Herschel, 30 August 1821, HRC H/L-0553.

28. Good, "John Herschel's Travels through the Alps," 285.

29. John Herschel to Mary Pitt Herschel, 15 September 1821, HRC H/L-0515.4.

30. John Herschel to Mary Pitt Herschel, 15 September 1821, HRC H/L-0515.4.

31. John Herschel to Gilbert Elliot, Second Earl of Minto, 16 June 1822, RS:HS 20.143; Good, "Herschel's Travels through the Alps," 286. This sketch is number 322 in the Graham Nash Collection according to Schaaf, *Tracings of Light*, but I have been unable to locate it in the collection of the Getty Museum.

32. John Herschel to Gilbert Elliot, Second Earl of Minto, 16 June 1822, RS:HS 20.143; Babbage, "Barometrical Observations"; John Herschel to William Herschel, 3 October 1821, HRC H/L-0554.

33. Good, "Astronomer Who Fell to Earth," 10.

34. Herschel diary, 30 November 1821, HRC container 16.8. See also the notice in *Philosophical Transactions of the Royal Society of London* 112 (1822): viii.

35. John Herschel to Charles Babbage, 4 April and 25 December 1820, RS:HS 2.133/20.87 and 2.151/20.107.

36. Herschel, "On Certain Remarkable Instances"; Herschel, "On the Rotation Impressed"; Herschel, "On the Reduction of Certain Classes"; John Herschel to Charles Babbage, 20 April 1820, RS:HS 2.134/20.91.

37. John Herschel to Charles Babbage, 2 December 1821, RS:HS 2.168/20.130.

38. See, for instance, Francis Baily to John Herschel, 4 January and 12 February 1822, RS:HS 3.45 and 3.48. For responses to these requests see Thomas Young to John Herschel, 7 February and 4 April 1822, RS:HS 18.321, 18.323, and 18.324.

39. Herschel diary, 18 June 1822, HRC container 16.9.

40. James Grahame to John Herschel, 15 July and 25 July 1822, RS:HS 8.309 and 8.310.

41. John Herschel to Mary Pitt Herschel, 14 August 1822, HRC H/L-0516.

42. Mary Pitt Herschel to John Herschel 13 August and 20 August 1822, HRC H/M-0621.4 and 0621.5; Thomas Beckwith to John Herschel, 23 August 1822, HRC H/M-0079.

43. John Herschel to James Grahame, 4 September 1822, RS:HS 19.29.

44. Herschel diary, 18 September 1822, HRC container 16.9.

CHAPTER 5: GRAND TOUR

Epigraph: John Herschel to Mary Pitt Herschel, 1 June 1824, HRC H/L-0517.10. Herschel is offering his own rendering of a stanza from Lord Byron's "Childe Harold's Pilgrimage": "O thou, Parnassus! whom I now survey, / Not in the frenzy of a dreamer's eye, / Not in the fabled landscape of a lay, / But soaring snow-clad through thy native sky."

1. On Herschel's inheritance, see Hingley, "Will of Sir William Herschel."

2. John Herschel to James South, 21 December 1823, RS:HS 20.171.

3. Hall, *All Scientists Now*, 30; John Herschel to James South, 8 January 1824, RS:HS 20.173.

4. For evidence that Lee was never an assistant, see Hollis, "Greenwich Assistants." John Herschel to James South, 15 February 1824, RS:HS 20.177.

5. Herschel travel journal, 4 April 1824, HRC container 22.5.

6. Rudwick, *Bursting the Limits of Time*, 465, 603, 618; Herschel travel journal, HRC container 22.4.

7. John Herschel to Mary Pitt Herschel, 7 April 1824, HRC H/L-0517.5.

8. For more on the actinometer and its use, see Voskulh, "Recreating Herschel's Actinometry."

9. Taylor, *Mathematical Practitioners*, 43; John Herschel to Mary Pitt Herschel, 19 April 1824, HRC H/L-0517.6; John Herschel to Charles Babbage, 11 April 1824 [finished 22 April], RS:HS 2.194.

10. John Herschel to Charles Babbage, 5 May 1824, RS:HS 20.180.

11. Szabados, "von Zach"; John Herschel to Charles Babbage, 5 May 1824, RS:HS 20.180; John Herschel travel journal, 24 April 1824, HRC container 22.5.

12. John Herschel to Mary Pitt Herschel, 26 April 1824, HRC H/L-0517.7; John Herschel to Charles Babbage, 5 May 1824, RS:HS 20.180.

13. John Herschel to Charles Babbage, 5 May 1824, RS:HS 20.180.

14. John Herschel to James Grahame, 2 May 1824, RS:HS 20.179.

15. John Herschel to Charles Babbage, 5 May 1824, RS:HS 20.180.

16. Herschel's poems included "To the Lark," "The Sailor's Departure," and "The Lament," which were published in Baillie, *Collection of Poems*, 81–83, 142, and 192–93. "The Lament" and "The Sailor's Departure" had originally been published in 1816. John Herschel to Mary Pitt Herschel, 11 May 1824, HRC H/L-0517.8.

17. Brewer, "Scientific Networks, Vesuvius and Politics."

18. Rudwick, *Worlds before Adam*, 209; Pyle, "Visions of Volcanoes."

19. Pyle, "Visions of Volcanoes," 13.

20. John Herschel to Mary Pitt Herschel, 8 June 1824, HRC H/L-0517.11.

21. Rudwick, *Worlds before Adam*, 106, 112; Lyell, *Principles of Geology*, vol. 1.

22. John Herschel to Mary Pitt Herschel, 17 June 1824, HRC H/L-0517.12; John Herschel travel journal, 4 June 1824, Herschel Family Papers, HRC container 22.6.

23. John Herschel to Mary Pitt Herschel, 20 June 1824, HRC H/L-0517.13.

24. John Herschel to William Watson, 16 July 1824, RS:HS 20.183.

25. John Herschel to Mary Pitt Herschel, 1 July 1824, HRC H/L-0517.15; John Herschel to William Watson, 16 July 1824, RS:HS 20.183; John Herschel to Mary Pitt Herschel, 27 June and 12 July 1824, HRC H/L-0517.14 and

L–0517.16.

26. Rudwick, *Worlds before Adam*, 97; Pyle, "Visions of Volcanoes," 3.

27. John Herschel to Charles Babbage, 11 July 1824, RS:HS 20.181.

28. John Herschel to Mary Pitt Herschel, 12 July 1824, HRC H/L–0517.16; Rudwick, *Worlds before Adam*, 127, 181.

29. John Herschel to Mary Pitt Herschel, 12 July 1824, HRC H/L–0517.16; John Herschel to Charles Babbage, 4 July 1824, RS:HS 20.181; John Herschel to Mary Pitt Herschel, 12 July and 30 July 1824, HRC H/L–0517.16 and L–0517.18; John Herschel to William Watson, 16 July 1824, RS:HS 20.183; John Herschel travel journal, 11 July 1824, HRC container 22.6.

30. John Herschel travel journal, 20–29 July 1824, HRC container 22.6; Blessington, *Conversations of Lord Byron*.

31. John Herschel to Mary [Baldwin], 6 August 1824, HRC H/L–0061; John Herschel to Mary Pitt Herschel, 14 August, 22 August, and 30 August, HRC H/L–0517.19, 0517.20, and 0517.21.

32. John Herschel to Mary Pitt Herschel, 9 September 1824, HRC H/L–0517.22.

33. John Herschel travel journal, 19 September 1824, HRC container 22.7; John Herschel to Charles Babbage, 3 October 1824, RS:HS 2.199.

34. John Herschel to James South, 21 September 1824, RS:HS 20.186.

35. John Herschel to Mary Pitt Herschel, 4 October and 11 October 1824, HRC H/L–0517.26 and 0517.27.

CHAPTER 6: A LONDON DISCONTENT

Epigraph: John Herschel to Charles Babbage, 12 February 1828, RS:HS 2.219.

1. For Gilbert's life and career, see Todd, *Beyond the Blaze*.

2. John Herschel to Francis Lunn, 5 January 1825, RS:HS 11.412.

3. For Herschel's work on hyposulforous acid, in which he discovered the "hypo" or "fixer" that would later be applied to photography, see "Herschel and Hypo," in Ross, *Nineteen-Century Attitudes*, 173–93. Herschel, "Light," 341.

4. Herschel, "Light," 367.

5. Herschel, "Light," 503, 515; John Herschel to William Whewell, 28 August 1826, RS:HS 20.237.

6. Translations of "Light" included *Traité de la lumière*, (2 vols.), trans. P. F. Verhulst and A. Quetelet (Paris, 1829–1833) and the German translation by J. E. Eduard Schmidt (Stuttgart and Tubingen, 1831). Herschel, "Light," 503 and 450.

7. John Herschel to Thomas Young, 8 November 1825, RS:HS 20.223.

8. John Herschel diary, 29 December 1825, HRC container 16.12.

9. John Herschel to James South, 14 November 1826, RS:HS B27.1; John Herschel to Charles Babbage, 17 November 1826, RS:HS 27.2.

10. John Herschel to Charles Babbage, 17 November 1826, RS:HS 27.2.

11. John Herschel to unknown, n.d., RS:HS B27.14.

12. John Herschel to Charles Babbage, RS:HS B27.9; John Herschel to Francis Baily, 25 November 1826, RS:HS B27.14.

13. Gleason, *Royal Society of London*, 106.

14. Charles Babbage to John Herschel, 16 December 1826, RS:HS 2.207; Georgiana Babbage to John Herschel, 29 January 1827, B27.10.

15. John Herschel to Charles Babbage, 30 January 1827, RS:HS 2.208.

16. John Herschel to Caroline Herschel, 4 May 1827, HRC H/L-0572.1.

17. James Grahame to John Herschel, 1 February 1827, RS:HS 8.350; John Herschel to James Grahame, 11 February 1827, RS:HS 20.249.

18. Todd, *Beyond the Blaze*, 226–27; John Herschel to William Whewell, 24 November 1823, RS:HS 20.166.

19. Todd, *Beyond the Blaze*, 228.

20. John Herschel to Charles Babbage, 27 June 1827, RS:HS 2.211; John Herschel to Davies Gilbert, 27 June 1827, RS:HS 20.257.

21. John Herschel to Caroline Herschel, 18 April and 27 April 1826, HRC H/L-0570.1 and L-0571.1.

22. John Herschel, "Account of Some Observations."

23. John Herschel diary, 9 February 1827, HRC container 16.14; John Herschel to William Henry Smyth, 19 December 1828, RS:HS 20.266.

24. John Herschel to Francis Baily, 12 March 1828, RS:HS 3.74.

25. Herschel, "On the Parallax of the Fixed Stars"; John Herschel to Mary Pitt Herschel, 13 April 1828, HRC H/L-0063; John Herschel to Charles Babbage, 10 April 1828, RS:HS 2.225.

26. Todd, *Beyond the Blaze*, 237.

27. John Herschel to Charles Babbage, 22 December 1827, RS:HS 2.218; Davies Gilbert to John Herschel, 26 December 1827, RS:HS 8.112; John Herschel to Davies Gilbert, 28 December 1827, RS:HS 20.270.

28. Howse, "Britain's Board of Longitude," 405.

29. John Herschel to Henry Kater, 17 July 1828, RS:HS 21.19.

30. John Herschel to Thomas Young, 30 August 1828, RS:HS 18.345; John Herschel to Davies Gilbert, 1 September 1828, RS:HS 21.27.

31. Herschel, *Treatises*, note to "Sound," §320, 810; John Herschel to Henry Kater, 1 August 1828, RS:HS 21.25.

CHAPTER 7: STABILITY OF THE SYSTEM

1. Newton, *Mathematical Principles*, 504.

2. For an overview of these fears and an examination of conservative reactions, specifically among some of Herschel's Cambridge colleagues, see Ashworth, *Trinity Circle*, esp. chs. 1 and 2.

3. John Herschel to John William Whittaker, 2 July 1813, St John's College, Cambridge, Herschel's emphasis.

4. John Herschel to Mary Pitt Herschel, 8 June 1824 and 20 May 1824, HRC H/L-0517.11 and L-0517.9.

5. John Herschel to James Grahame, 18 September 1824, RS:HS 20:185.

6. John Herschel to Mary Pitt Herschel, 3 September 1826, HRC H/L-0518.2; Rudwick, *Worlds before Adam*, 89, 127, 209; Rudwick, *Bursting the Limits of Time*, 618; Scrope, *Considerations on Volcanoes*; Daubeny, *Description of Active and Extinct Volcanoes*.

7. John Herschel to Mary Pitt Herschel, 11 September 1826, HRC H/L-0518.3; John Herschel to William Whewell, 17 October 1826, RS:HS 20.240; John Herschel to Charles Babbage, 25 September 1826, RS:HS 2.206.

8. Herschel, "On the Astronomical Causes."

9. John Herschel to James Grahame, 18 September 1824, RS:HS 20:185; Alborn, "Business of Induction," 108.

10. Babbage published a short account in the first volume of the Astronomical Society's *Memoirs*: Babbage, "Note Respecting the Application of Machinery." On the response to news of Babbage's device during his travels, see for example John Herschel to Charles Babbage, 11 April and 3 October 1824, RS:HS 2.194 & 2.199.

11. Herschel, "Calculating Machinery."

12. John Herschel to Miles Bland, 10 March 1821, RS:HS 4.154; Miles Bland to John Herschel, 11 June 1822, RS:HS 4.155; and John Herschel to Miles Bland, 12 June 1822, RS:HS 19.19; Hutchins, *British University Observatories*, 33.

13. John Herschel to Thomas W. Hornbuckle, 13 October 1826, RS:HS 20.283.

14. John Herschel to William Whewell, 17 October 1826, RS:HS 20.240.

15. John Herschel to Richard Gwatkin, 26 December 1827, RS:HS 20.269.

16. James Grahame to John Herschel, 6 October 1827, RS:HS 8.359.

17. James Grahame to John Herschel, 24 April 1826, RS:HS 8.338; Caroline Herschel to John Herschel, 1 February 1826, in Herschel, *Memoir and Correspondence*, 197; James Grahame to John Herschel, 8 October 1826, RS:HS 8.343.

18. John Herschel to James Grahame, 1 July 1825, RS:HS 20.216, Herschel's emphasis.

19. For more on the Stewart family, see Winterburn, "John Herschel," in Case and Verburgt, *Cambridge Companion to John Herschel*.

20. John Herschel diary, 24 March 1827, 21 and 25 September 1828, HRC container 16.15 and 16.16.

21. John Herschel to Emilia Stewart, 4 December 1828, private collection of William Herschel-Shorland (hereafter JHS) 2.24.

22. John Herschel to James Grahame, 27 October 1828, RS:HS 8.375; John Herschel diary, 9 November 1829, HRC container 16.16.

23. Turner, "Victorian Crisis of Faith," 20–21.

24. James Grahame to John Herschel, 11 November 1828, RS:HS 8.377.

25. For Charles Wesley, Jr., see "Anecdotes of the Late Charles Wesley"; and Lloyd, "Charles Wesley, Junior." For the invitation to Slough, see John Herschel to Charles Wesley, 26 September 1828, RS:HS 18.142.

26. John Herschel to James Grahame, 19 November 1828, RS:HS 8.378; James Grahame to John Herschel, 25 November 1828, RS:HS 8.379. Wesley must not have been too unhappy with the sudden change in his domestic situation, as he spent the rest of his life in the company of the ladies who had taken him away from Mrs. Mortimer's, doing little more than playing music at all hours of the day. Eventually the Stewarts reestablished contact with their friend, and Emilia sent Herschel to call on Wesley and ask him to visit them. John Herschel to Emilia Stewart, 28 November 1828, JHS 2.23.

27. John Herschel to Emilia Stewart, 27 November 1828, JHS 2.21.

28. James Grahame to Peter Stewart, 29 November 1828, JHS 2.26.

29. John Herschel to Emilia Stewart, 4 December 1828, JHS 2.24.

30. Margaret Brodie Stewart to John Herschel, n.d., JHS 2.27; John Herschel diary, 7 and 11 December 1828, HRC container 16.15.

31. John Herschel to Margaret Brodie Stewart, 24 December 1828, JHS 2.18.

CHAPTER 8: REFORM DEFEATED

Epigraph: John Herschel to Charles Babbage, 26 November 1830, RS:HS 2.257.

1. Mary Baldwin to Caroline Herschel, 3 March 1829, JHS 503.

2. John Herschel to Mary Pitt Herschel, 5 March 1829, HRC H/L-0521.1; John Herschel diary, 10 March 1829, HRC container 16.16; John Herschel to James Grahame, 27 March 1829, RS:HS 8.385.

3. John Herschel diary, 2–4 and 15 April 1829, HRC container 16.16; John Herschel to Mary Pitt Herschel, 17 April 1829, HRC H/L-0521.2.

4. John Herschel diary, 12 and 13 May 1829, HRC container 16.16; John Herschel to Mary Pitt Herschel, 14 May 1829, HRC H/L-0521.3; Margaret Herschel to Mary Pitt Herschel, 15 May 1829, HRC H/L-0521.4.

5. John Herschel to William Whewell, 17 May 1829, RS:HS 21.47.

6. John Herschel to Caroline Herschel, 5 February 1829, JHS 502; John Herschel to Mary Pitt Herschel, 24 June 1829 and 7 June 1829, HRC H/L-0521.9 and 0521.7.

7. John Herschel to Mary Pitt Herschel, 2 July 1829, HRC H/L-0521.10; James Grahame to John Herschel, 29 June 1829, RS:HS 8.386.

8. John Herschel to Mary Pitt Herschel, 6 July 1829, HRC H/L-0521.11.

9. See Secord, *Visions of Science*, 14–21.

10. John Herschel to Charles Babbage, 14 July 1816, RS:HS 2.64, Her-

schel's emphasis. For details of the *Metropolitana* arrangements, see John Herschel to Francis Lunn, 5 January 1825, RS:HS 11.411. On the *Metropolitana* and other contemporary encyclopedias, see Yeo, "Reading Encyclopedias." For details on Lardner, see Peckham, "Lardner's 'Cabinet Cyclopedia'"; and Hays, "Rise and Fall of Dionysius Lardner."

11. Value calculated with "Five Ways to Compute the Relative Value of a UK Pound Amount, 1270 to Present," MeasuringWorth.com, 2023. Secord, *Victorian Sensation*, 50–51; Dionysius Lardner to John Herschel, 18 July 1828, RS:HS 11.106.

12. John Herschel to Dionysius Lardner, 25 July 1828, RS:HS 21.23.

13. Dionysius Lardner to John Herschel, 28 July 1828, RS:HS 11.108; John Herschel to Dionysius Lardner, 1 August 1828, RS:HS 21.24.

14. John Herschel to Dionysius Lardner, 16 February 1829, RS:HS 11.111.

15. For Scott's contribution to the *Cyclopedia* see Scott, *History of Scotland*. Dionysius Lardner to John Herschel, 19 February 1829, RS:HS 11.113; John Herschel to Dionysius Lardner, 28 February 1829, RS:HS 11.114.

16. Charles Babbage to John Herschel, 6 May 1829, RS:HS 2.239. In this letter Babbage also gives Herschel marriage advice: *"Never allow a thought in your mind which is not shared with your wife"* (emphasis in original). Charles Babbage to John Herschel, 11 December 1829, RS:HS 2.241. Babbage adds in this letter that after Herschel's marriage, "Everybody asks for you but nobody sees you. We all regret it: but we ought not for it is a *good* sign" (emphasis in original).

17. John Herschel to Charles Babbage, 15 December 1829, RS:HS 2.242.

18. Herschel, "Sound," 810n.

19. John Herschel to Davies Gilbert, 24 November 1829, RS:HS 8.123.

20. Herschel, "Sound," 810n.

21. John Herschel diary, 3 February 1830, HRC container 16.17; John Herschel to Charles Babbage, 28 February 1830, RS:HS 2.243.

22. Charles Babbage to John Herschel, 4 March 1830, RS:HS 2.244 and 19 March 1830, RS:HS 2.246; Babbage, *Reflections on the Decline of Science*, vii–viii.

23. John Herschel to Charles Babbage, [8] March 1830, RS:HS 2.245.

24. Charles Babbage to John Herschel, 19 March 1830, RS:HS 2.246. For more on the content and reception of Babbage's book, see "The Economy of Intelligence: Charles Babbage's *Reflections on the Decline of Science in England*" in Secord, *Visions of Science*, 52–79.

25. John Herschel diary, 9 December 1829, HRC container 16.16; John Herschel to Margaret Herschel, 15 December 1829 and 3 January 1830, JHS 504c and 505a; John Herschel to Thomas Smart Hughes, 15 January 1830, RS:HS 21.55.

26. Secord, *Victorian Sensation*, 42–43.

27. Dionysius Lardner to John Herschel, 19 January 1830, RS:HS 11.115; John Herschel to Dionysius Lardner, 1 February 1830, RS:HS 11.116.

28. Somerville, *Mechanism of the Heavens*; Gurney, *Memoir of the Life of Thomas Young*.

29. Charles Babbage to John Herschel, 27 March 1830, RS:HS 2.251.

30. John Herschel diary, 1 April 1830, HRC container 16.17; John Herschel to Emilia Stewart, 5 April 1830, JHS 508.

31. Snyder, *Philosophical Breakfast Club*, 134. For more on Whewell's response to Herschel see Yeo, "Reviewing Herschel's *Discourse*."

32. John Herschel to Hudson Gurney, 25 June 1830.

33. Herschel, *Preliminary Discourse*, xxviii.

34. John Herschel to Margaret Herschel, 23 and 28 July 1830, JHS 513 and 514.

35. Herschel, *Preliminary Discourse*, 7.

36. Dionysius Lardner to John Herschel, 21 September 1830, RS:HS 11.117.

37. John Herschel to Dionysius Lardner, [September], RS:HS 11.118. His degree from Cambridge was a different matter. It was not "a mere honourary distinction," he explained to Lardner, but rather "a legal qualification," and so formed "part of the designation of those who bear it." With these considerations in mind, he requested his name appear on the cover page of the *Discourse* as "John Frederick William Herschel Esq. AM, Late Fellow of St. John's College Cambridge."

38. See, for instance, *The Standard*, 13 October 1830, 3. By the end of the month, notices for the publication of the *Preliminary Discourse* were also appearing in this newspaper and *The Morning Post*.

39. Quoted in Gleason, *Royal Society of London*, 277–78; "To the Editor of the Times," *The Times*, 25 October 1830, page 3.

40. John Herschel to William Henry Fitton, 18 October 1830, RS:HS 21.1.9.

41. John Herschel to William Henry Fitton, 18 October 1830, RS:HS 21.1.9.

42. John Herschel to William Henry Fitton, 18 October 1830, RS:HS 21.1.9.

43. Herschel diary, 21 and 22 October 1830, HRC container 16.17; John Herschel to Margaret Herschel, 27 October 1830, JHS 517c.

44. These notices appeared in *The Times*, 25 November, 3, and 29 November, 4, and 30 November 1830, 4. As first published it was signed by sixty-three fellows; by November 29, the revised list contained eighty names.

45. John Herschel to Charles Babbage, 26 November 1830, RS:HS 2.257.

46. John Herschel to Francis Beaufort, 26 November 1830, RS:HS 25.1.13.

47. *The Times*, 29 November 1830, 4; Snyder, *Philosophical Breakfast Club*, 139.

CHAPTER 9: THE INVENTION OF SCIENCE

Epigraph: Dionysius Lardner to John Herschel, 2 October 1830, RS:HS 11.119.

1. Herschel, "Observations of Nebulae"; and Herschel, *Treatise on Astronomy*. For details on Herschel's trip to the Cape, see Ruskin, *Cape Results*; and Evans, *Herschel at the Cape*. See also Steve Ruskin, "Stargazer at World's End: John Herschel at the Cape, 1833–1838," in Case and Verburgt, *Cambridge Companion to John Herschel*.

2. Browne, *Charles Darwin*, quoted on 128. For the influence of the *Discourse* on Darwin, see Phillip R. Sloan, "The Making of a Philosophical Naturalist" and C. Kenneth Waters, "The Arguments in the *Origin of Species*," both in Hodge and Radick, *Cambridge Companion to Darwin*, 21–43 and 120–43; Ruse, "Darwin's Debt to Philosophy"; and Pence, "Herschel and Darwin."

3. Browne, *Charles Darwin*, 418; Whewell, Review of *A Preliminary Discourse*, 377 and 398; Faraday quoted in Yeo, "Reviewing Herschel's *Discourse*," 541; Mackintosh quoted in Secord, *Visions of Science*, 102.

4. *Athenaeum*, review of *A Preliminary Discourse*, 38–39; Whewell, Review of *A Preliminary Discourse*, 389, 406.

5. Review of Herschel's *Discourse*, 763, 858; Secord, *Visions of Science*, quoted on 100–101.

6. Herschel, *Preliminary Discourse*, 4, 7 (hereafter cited in text as *PD*).

7. See Cunningham and Williams, "De-centring the 'Big Picture'"; Harrison, *Territories of Science and Religion*, esp. 145–59; and Herschel, *Preliminary Discourse*, 18.

8. Herschel, *Preliminary Discourse*, 20, 22.

9. Herschel, *Preliminary Discourse*, 25.

10. Herschel, *Preliminary Discourse*, 43.

11. Herschel, *Preliminary Discourse*, 59, 69, 70.

12. Herschel, *Preliminary Discourse*, 73–74.

13. John Herschel to Charles Babbage, 9 December 1830, RS:HS 2.258.

14. Herschel, *Preliminary Discourse*, 77.

15. Herschel, *Preliminary Discourse*, 92.

16. Herschel, *Preliminary Discourse*, 95, 96, 97.

17. Herschel, *Preliminary Discourse*, 99–100.

18. Herschel, *Preliminary Discourse*, 103.

19. Herschel, *Preliminary Discourse*, 148.

20. Herschel, *Preliminary Discourse*, 164.

21. Herschel, *Preliminary Discourse*, 171.

22. Herschel, *Preliminary Discourse*, 173.

23. Herschel, *Preliminary Discourse*, 190.

24. Herschel, *Preliminary Discourse*, 219, 223, 230.

25. Herschel, *Preliminary Discourse*, 245.

26. Herschel, *Preliminary Discourse*, 257, 263, 259.

27. Herschel, *Preliminary Discourse*, 279–80.

28. On Herschel's unification of terrestrial and celestial phenomena, see Good, "Astronomer who Fell to Earth"; Herschel, *Preliminary Discourse*, 282, 285, 289.

29. Herschel, *Preliminary Discourse*, 293, 299, 306.

30. Herschel, *Preliminary Discourse*, 350–51.

31. Herschel, *Preliminary Discourse*, 360–61

32. For the formation of the British Association for the Advancement of Science, see Morrell and Thackray, *Gentlemen of Science*. On the origin and evolution of the term "scientist," see Ross, "Scientist: the Story of a Word," in Ross, *Nineteenth-Century Attitudes*, 1–39. For the changing relationship between science and theology after the *Preliminary Discourse*, see Harrison, *Territories of Science and Religion*, 159–80.

33. Brown, *Charles Darwin*, quote on 328–29.

EPILOGUE

1. Case, *Making Stars Physical*, 131, 133; Steinicke, *Observing and Cataloguing Nebulae and Star Clusters*, 53; Airy quoted in Case, *Making Stars Physical*, 125.

2. Herschel, "Observations of Nebulae and Clusters of Stars"; Case, *Making Stars Physical*, 129.

3. Case, *Making Stars Physical*, 133.

4. Morrell and Thackray, *Gentlemen of Science*, 80–81.

5. Quoted in Ruskin, *Herschel's Cape Voyage*, 48–49.

6. Warner and Rourke, *Flora Herscheliana*.

7. Desmond, *Darwin's Sacred Cause*, 106.

8. Gleason, *Royal Society of London*, 379, 304, 399. For a recent study on the role of Lubbock in reforming the Royal Society, see Fullick, "Science without a Head?"

BIBLIOGRAPHY

Alborn, Timothy. "The Business of Induction: Industry and Genius in the Language of British Scientific Reform." *History of Science* 34, no. 1 (1996): 91–121.

"Anecdotes of the Late Charles Wesley, Esq." *Wesleyan Methodist Magazine* 13 (1834): 514–18.

Ashworth, William J. "The Calculating Eye: Baily, Herschel, and the Business of Astronomy." *British Journal for the History of Science* 27, no. 4 (1994): 409–41.

Ashworth, William J. "Memory, Efficiency, and Symbolic Analysis: Charles Babbage, John Herschel, and the Industrial Mind." *Isis* 87, no. 4 (1996): 629–53.

Ashworth, William J. *The Trinity Circle: Anxiety, Intelligence, and Knowledge Creation in the Nineteenth Century.* Pittsburgh: University of Pittsburgh Press, 2021.

Athenaeum. Review of *A Preliminary Discourse on the Study of Natural Philosophy* by J. F. W. Herschel. *Athenaeum* 167 (8 January 1831): 38–39.

Babbage, Charles. "Barometrical Observations Made at the Fall of Staubbach." *Edinburgh Philosophical Journal* 6 (1822): 224–27.

Babbage, Charles. "A Note Respecting the Application of Machinery to the Calculation of Astronomical Tables." *Memoirs of the Astronomical Society* 1 (1822): 309.

Babbage, Charles. *Reflections on the Decline of Science in England.* London: B. Fellowes, 1830.

Ball, W. W. Rouse. *A History of Mathematics at Cambridge.* Mansfield Center, CT: Martino, 2004.

Bayly, Christopher. *The Birth of the Modern World: 1780–1914*. Oxford: Blackwell, 2004.

Becher, Harvey. "William Whewell and Cambridge Mathematics." *Historical Studies in the Physical Sciences* 11, no. 1 (1981): 1–48.

Becher, Harvey W. "Radicals, Whigs, and Conservatives: the Middle and Lower Classes in the Analytical Revolution in the Age of Aristocracy." *British Journal for the History of Science* 28, no. 4 (1995): 405–26.

Blessington, Margaret. *Conversations of Lord Byron with the Countess of Blessington*. London: Henry Colburn, 1834.

Boase, Frederic. *Modern English Biography*. Vol. 1. New York: Barnes and Noble, 1965.

Bolt, Marvin. "John Herschel's Natural Philosophy: On the Knowing of Nature and the Nature of Knowing in Early Nineteenth-Century Britain." PhD diss., University of Notre Dame, 1998.

Bord, Joe. *Science and Whig Manners: Science and Political Style in Britain, c. 1790–1850*. Basingstoke, UK: Palgrave Macmillan, 2009.

Boyd, Hilton. *A Mad, Bad, Dangerous People? England, 1783–1846*. Oxford: Clarendon, 2011.

Brewer, John. "Scientific Networks, Vesuvius and Politics: the Case of Teodoro Monticelli in Naples, 1790–1845." *Incontri* 30 (2019): 54–67.

Browne, Janet. *Charles Darwin: A Biography*. Princeton, NJ: Princeton University Press, 1996.

Buttmann, Günther. *The Shadow of the Telescope: A Biography of John Herschel*. Guildford: Lutterworth Press, 1974.

Cagnoli, Antonio. *Memoir on a New and Certain Method of Ascertaining the Figure of the Earth by Means of Occultations of the Fixed Stars*. Translated by Francis Baily. London: Taylor, 1819.

Case, Stephen. *Making Stars Physical: The Astronomy of Sir John Herschel*. Pittsburgh: University of Pittsburgh Press, 2018.

Case, Stephen, and Lukas M. Verburgt, eds. *The Cambridge Companion to John Herschel*. Cambridge: Cambridge University Press, 2024.

"Clark, Edward Daniel." In *Complete Dictionary of Scientific Biography*, vol. 3, 290–92. Detroit: Charles Scribner's Sons, 2008.

Cobb, Aaron D. "Inductivism in Practice: Experiment in John Herschel's Philosophy of Science." *HOPOS: The Journal of the International Society for the History of the Philosophy of Science* 2 (2012): 21–54.

Cobb, Aaron D. "Is John F. W. Herschel an Inductivist about Hypothetical Inquiry?" *Perspectives on Science* 20, no. 4 (2012): 409–39.

Colley, Linda. *Britons: Forging the Nation, 1707–1837*. New Haven, CT: Yale University Press, 1992.

Craik, Alex D. D. "Calculus and Analysis in Early 19th-Century Britain: The Work of William Wallace." *Historia Mathematica* 26 (1999): 239–67.

Craik, Alex D. D. "Polylogarithms, Functional Equations and More: the Elusive Essays of William Spence (1777–1815)." *Historia Mathematica* 40 (2013): 386–422.

Crosland, Maurice. *Science under Control: The French Academy of Sciences, 1795–1914.* Cambridge: Cambridge University Press, 1992.

Crosland, Maurice. *The Society of Arcueil: A View of French Science at the Time of Napoleon I.* Cambridge, MA: Harvard University Press, 1967.

Crosland, Maurice, and Crosbie Smith. "The Transmission of Physics from France to Britain: 1800–1840." *Historical Studies in the Physical Sciences* 9 (1978): 1–61.

Crowe, Michael J. *The Extraterrestrial Life Debate, 1750–1900.* Mineola, NY: Dover, 1999.

"Cumming, James." In *Complete Dictionary of Scientific Biography*, vol. 3, 497. Detroit: Charles Scribner's Sons, 2008.

Cunningham, Andrew. "Getting the Game Right: Some Plain Words on the Identity and Invention of Science." *Studies in History and Philosophy of Science* 19 (1988): 365–89.

Cunningham, Andrew, and Perry Williams. "De-centring the 'Big Picture': 'The Origins of Modern Science' and the Modern Origins of Science." *British Journal for the History of Science* 26, no. 4 (1993): 407–32.

Daubeny, Charles. *A Description of Active and Extinct Volcanoes; with Remarks on their Origin, their Chemical Phenomena, and the Character of their Products, as Determined by the Condition of the Earth during the Period of their Formation.* London: Phillips, 1826.

De Morgan, August. "Account of a Correspondence between Mr. George Barrett and Mr. Francis Baily." *The Assurance Magazine, and Journal of the Institute of Actuaries* 4, no. 3 (1854): 185–99.

Desmond, Adrian J., and James R. Moore. *Darwin's Sacred Cause: How a Hatred of Slavery Shaped Darwin's Views on Human Evolution.* Boston: Houghton Mifflin Harcourt, 2009.

Dettelbach, Michael. "Humboldtian Science." In *Cultures of Natural History*, edited by Nicholas Jardine and James A. Secord, 287–304. Cambridge: Cambridge University Press, 1996.

Dreyer, J. L. E., and H. H. Turner, eds. *History of the Royal Astronomical Society 1820–1920.* 1923. Reprint, Palo Alto, CA: Blackwell Scientific, 1987.

Dubbey, J. M. "The Introduction of the Differential Notation to Great Britain." *Annals of Science* 19 (1963): 37–48.

Edgeworth, Maria. *Letters from England, 1813–1844.* Oxford: Oxford University Press, 1971.

Edwards, A. W. F. "Bromhead, Sir Edward Thomas French, Second Baronet (1789–1855)." In *Oxford Dictionary of National Biography*, 818. Oxford: Oxford University Press, 2023.

Enros, Philip C. "The Analytical Society (1812–1813): Precursor of the Renewal of Cambridge Mathematics." *Historia Mathematica* 10, no. 1 (1983): 24–47.

Evans, David S., Terence J. Deeming, Betty Hall Evans, and Stephen Goldfarb, eds. *Herschel at the Cape: Diaries and Correspondence of Sir John Herschel, 1834–1838.* Austin: University of Texas Press, 1969.

Evans, Eric. *The Forging of the Modern State: Early Industrial Britain, 1783–1870.* London: Longman, 1983.

Fisch, Menachem. "Babbage's Two Lives." *British Journal for the History of Science* 47, no. 1 (2014): 95–118.

Fisch, Menachem. *Creatively Undecided: Toward a History and Philosophy of Scientific Agency.* Chicago: University of Chicago Press, 2017.

Fisch, Menachem. "'The Emergency Which Has Arrived': The Problematic History of Nineteenth-Century British Algebra—A Programmatic Outline." *British Journal for the History of Science* 27, no. 3 (1994): 247–76.

Fox, Robert. *The Savant and the State: Science and Cultural Polities in Nineteenth-Century France.* Baltimore, MD: Johns Hopkins University Press, 2012.

Fullick, Timothy Grenville. "Science without a Head? John William Lubbock and the Leadership of English Science in the 1830s." PhD diss., University of Kent, 2021.

Gascoigne, John. *Science in the Service of Empire: Joseph Banks, the British State and the Uses of Science in the Age of Revolution.* Cambridge: Cambridge University Press, 1998.

Gilbert, L. F. "The Election to the Presidency of the Royal Society in 1820." *Notes and Records of the Royal Society of London* 11, no. 2 (1955): 256–76.

Gillen, Mollie. *Assassination of the Prime Minister: The Shocking Death of Spencer Perceval.* London: Sidgwick and Jackson, 1972.

Gleason, Mary Louise. *The Royal Society of London: Years of Reform, 1827–1847.* New York: Garland, 1991.

Golinski, Jan. *Science as Public Culture: Chemistry and Enlightenment in Britain, 1760–1820.* Cambridge: Cambridge University Press, 1999.

Good, Gregory. "The Astronomer Who Fell to Earth: Or, How the Copernican Revolution Was Completed in the Alps." *Physis* 56, no. 1–2 (2021): 7–20.

Good, Gregory. "John Herschel's Optical Researches and the Development of His Ideas on Method and Causality." *Studies in History and Philosophy of Science* 18 (1987): 1–41.

Good, Gregory. "John Herschel's Travels through the Alps to the Cosmos in the 1820s." *Viaggiatori* 1, no. 1 (2018): 288–91.

Good, Gregory. "A Shift of View: Meteorology in John Herschel's Terrestrial Physics." In *Intimate Universality: Local and Global Themes in the History of Weather and Climate,* 35–67. New York: Science History, 2006.

Grahame, James. *The History of the United States of North America: From the Plantation of the British Colonies till Their Assumption of National Independence.* 2nd ed. 1845. Reprint, Freeport, NY: Books for Libraries Press, 1971.

Gunning, Henry. *Reminiscences of the University, Town, and County of Cambridge from the Year 1780.* Vol. 2. London: George Bell, 1854.

Gurney, Hudson. *Memoir of the Life of Thomas Young.* London: John & Arthur Arch, 1831.

Hahn, Roger. *Pierre Simon Laplace, 1749–1827: A Determined Scientist.* Cambridge, MA: Harvard University Press, 2005.

Hall, Marie Boas. *All Scientists Now: The Royal Society in the Nineteenth Century.* Cambridge: Cambridge University Press, 1984.

Harrison, Peter. *The Territories of Science and Religion.* Chicago: University of Chicago Press, 2017.

Hays, J. N. "The Rise and Fall of Dionysius Lardner." *Annals of Science* 38 (1981): 527–42.

Heilbron, J. L. "A Mathematician's Mutiny, with Morals." In *World Changes: Thomas Kuhn and the Nature of Science*, edited by Paul Horwich, 81–129. Cambridge, MA: MIT Press, 1993.

Herrmann, Luke. "The Drawings by Sir Joshua Reynolds in the Herschel Album." *Burlington Magazine* 110, no. 789 (1968): 650–58.

Herschel, John. "Additional Facts Relative to the Hyposulphurous Acid." *Edinburgh Philosophical Journal* 1 (1819): 396–400.

Herschel, John. "Address of the Society Explanatory of Their Views and Objects." *Memoirs of the Royal Astronomical Society* 1 (1822): 1–7.

Herschel, John. "Account of Some Observations Made with a 20-Feet Reflecting Telescope: Comprehending, 1. Descriptions and Approximate Places of 321 New Double and Triple Stars. 2. Observations of the Second Comet of 1825. 3. An Account of the Actual State of the Great Nebula in Orion, Compared with Those of Former Astronomers. 4. Observations of the Nebula in the Girdle of Andromeda." *Astronomical Society Memoirs* 2 (1826): 459–97.

Herschel, John. "An Address Delivered on the Occasion of the Distribution of the Honorary Medals on April 11, 1827, to Francis Baily, Esq., Lieutenant W. S. Stratford, R.N., and Colonel Mark Beaufoy." *Philosophical Magazine* 2 (1827): 455–66.

Herschel, John. "Analytical Formulae for the Tangent, Cotangent, &c." *Journal of Natural Philosophy, Chemistry, and the Arts* 31 (1812): 133–36.

Herschel, John. *A Collection of Examples of the Applications of the Calculus of Finite Differences.* Cambridge: J. Smith, 1820.

Herschel, John. "Consideration of Various Points of Analysis." *Philosophical Transactions of the Royal Society of London* 104 (1814): 440–68.

Herschel, John. "[Letter on] Calculating Machinery." *London Times*, 19 August 1828.

Herschel, John. "Light." *Encyclopaedia Metropolitana*, 2nd Division: Mixed Sciences, vol. 2, 341–586. London: Baldwin and Cradock, 1830.

Herschel, John. "Observations of Nebulae and Clusters of Stars, Made at Slough, with a Twenty-Feet Reflector between the Years 1825 and 1833." *Philosophical Transactions of the Royal Society* 123 (1833): 359–505.

Herschel, John. "On a Remarkable Application of Cotes's Theorem." *Philosophical Transactions of the Royal Society of London* 103 (1813): 8–26.

Herschel, John. "On Certain Optical Phenomena Exhibited by Mother-of-Pearl, Depending on Its Internal Structure." *Edinburgh Philosophical Journal* 2 (1820): 114–21.

Herschel, John. "On Certain Remarkable Instances of Deviation from Newton's Scale in the Tints Developed by Crystals with One Axis of Double Refraction on Exposure to Polarized Light." *Cambridge Philosophical Society Transactions* 1 (1822): 21–42.

Herschel, John. "On Equations of Differences and Their Application to the Determination of Functions from Given Conditions." In *Memoirs of the Analytical Society*, 65–114. Cambridge: J. Smith, 1813.

Herschel, John. "On the Aberrations of Compound Lenses and Object-Glasses." *Philosophical Transactions of the Royal Society of London* 111 (1822): 222–67.

Herschel, John. "On the Action of Crystallized Bodies on Homogeneous Light, and on the Causes of the Deviation of Newton's Scale in the Tints Which Many of Them Develope on Exposure to a Polarised Ray." *Philosophical Transactions of the Royal Society of London* 110 (1820): 45–100.

Herschel, John. "On the Application of a New Mode of Analysis to the Theory and Summation of Certain Extensive Classes of Series." *Edinburgh Philosophical Journal* 2 (1820): 23–33.

Herschel, John. "On the Astronomical Causes which May Influence Geological Phenomena." *Geological Society Transactions* 3 (1835): 293–99.

Herschel, John. "On the Development of Exponential Functions; Together with Several New Theorems relating to Finite Differences." *Philosophical Transactions of the Royal Society of London* 106 (1816): 25–45.

Herschel, John. "On the Hyposulphurous Acid and Its Compounds." *Edinburgh Philosophical Journal* 1 (1819): 8–29.

Herschel, John. "On the Parallax of the Fixed Stars." *Philosophical Transactions of the Royal Society of London* 116 (1826): 266–80.

Herschel, John. "On the Reduction of Certain Classes of Functional Equations to Equations of Finite Differences." *Cambridge Philosophical Society Transactions* 1 (1822): 77–87.

Herschel, John. "On the Rotation Impressed by Plates of Rock Crystal on the

Planes of Polarization of the Rays of Light, as Connected with Certain Peculiarities in Its Crystallization." *Cambridge Philosophical Society Transactions* 1 (1822): 43–52.

Herschel, John. "On Trigonometrical Series, Particularly Those Whose Terms Are Multiplied by the Tangents, Cotangents, Secants, &c. of Quantities in Arithmetic Progression, Together with Some Singular Transformations, with Notes relating to a Variety of Subjects Connected with the Preceding Memoir." In *Memoirs of the Analytical Society*, 33–64. Cambridge: J. Smith, 1813.

Herschel, John. *A Preliminary Discourse on the Study of Natural Philosophy.* 1830. Reprint, Chicago: University of Chicago Press, 1987.

Herschel, John. "Some Additional Facts relating to the Habitudes of the Hyposulphurous Acid and Its Union with Metallic Oxides." *Edinburgh Philosophical Journal* 2 (1820): 154–56.

Herschel, John. "Sound." *Encyclopaedia Metropolitana*, 2nd Division: Mixed Sciences, vol. 2, 747–824. London: Baldwin and Cradock, 1830.

Herschel, John. "To the Lark," "The Sailor's Departure," and "The Lament." In *A Collection of Poems, Chiefly Manuscript, and from Living Authors*, edited by Joanna Baillie, 81–83, 142, 192–93. London: Longman, 1823.

Herschel, John. *A Treatise on Astronomy.* London: Longman, Rees, Orme, Brown, Green & Longman, 1833.

Herschel, John. *Treatises: Physical Astronomy, Light, and Sound.* London: Richard Griffon, n.d.

Herschel, John. "Trigonometrical Formulae for Sines and Cosines, &c." *Journal of Natural Philosophy, Chemistry, and the Arts* 32 (1812): 13–16.

Herschel, John, and James South. "Observations of the Apparent Distances and Positions of 380 Double and Triple Stars, Made in the Years 1821, 1822, and 1823, and Compared with Those of Other Astronomers; Together with an Account of Such Changes as Appear to Have Taken Place in Them since Their First Discovery. Also a Description of a Five-Feet Equatorial Instrument Employed in the Observations." *Philosophical Transactions of the Royal Society of London* 114 (1824): 1–59.

Herschel, Mrs. John. *Memoir and Correspondence of Caroline Herschel.* London: John Murray, 1879.

Herschel, William. *The Scientific Papers of Sir William Herschel.* London: Royal Society and the Royal Astronomical Society, 1912.

Hevley, Bruce. "The Heroic Science of Glacier Motion." *Osiris* 11 (1996): 66–86.

Hingley, P. D. "The Will of Sir William Herschel." *Astronomy & Geophysics* 39 (1998): 3–7.

Hodge, M. J. S., and Gregory Radick, eds. *The Cambridge Companion to Darwin.* 2nd ed. Cambridge: Cambridge University Press, 2009.

Hollis, H. P. "The Greenwich Assistants during 250 Years." *The Observatory*, no. 619 (1925): 388–98.

Hoskin, Michael. *The Construction of the Heavens: William Herschel's Cosmology*. Cambridge: Cambridge University Press, 2012.

Hoskin, Michael. *Discoverers of the Universe: William and Caroline Herschel*. Princeton, NJ: Princeton University Press, 2011.

Hoskin, Michael. "Mary Herschel's Fortune: Origins and Impact." *Journal for the History of Astronomy* 41 (2010): 213–23.

Hoskin, Michael. "William Herschel's Agenda for His Son John." *Journal for the History of Astronomy* 43, no. 4 (2012): 439–54.

Howse, Derek. "Britain's Board of Longitude: The Finances, 1714–1828." *Mariner's Mirror* 84, no. 4 (1998): 400–417.

Hutchins, Roger. *British University Observatories 1772–1939*. Aldershot, UK: Ashgate, 2008.

Kaalund, Nanna Katrine Lüders. "A Frosty Disagreement: John Tyndall, James David Forbes, and the Early Formation of the X-Club." *Annals of Science* 74, no. 4 (2017): 282–98.

Koppleman, Elaine. "The Calculus of Operations and the Rise of Abstract Algebra." *Archive for the History of Exact Sciences* 8, no. 3 (1971): 155–242.

Lacroix, Silvestre François. *An Elementary Treatise on the Differential and Integral Calculus*. Translated by John Herschel, Charles Babbage, and George Peacock. Cambridge: J. Smith, 1816.

Lal Jain, Chaman. "Methodology and Epistemology: An Examination of Sir John Frederick William Herschel's Philosophy of Science with Reference to His Theory of Knowledge." PhD diss., Indiana University, 1975.

Linehan, Peter, ed. *St John's College Cambridge: A History*. Woodbridge, Suffolk, UK: Boydall Press, 2011.

Linklater, Andro. *Why Spencer Perceval Had to Die: The Assassination of a British Prime Minister*. New York: Walker, 2012.

Lloyd, Gareth. "Charles Wesley, Junior: Prodigal Child, Unfulfilled Adult." *Proceedings of the Charles Wesley Society* 5 (1998): 23–35.

Lubbock, Contance. *The Herschel Chronicle: The Life-Story of William Herschel and His Sister Caroline Herschel*. Cambridge: Cambridge University Press, 1933.

Lubenow, William C. *"Only Connect": Learned Societies in Nineteenth-Century Britain*. Woodbridge, Suffolk, UK: Boydell Press, 2015.

Luminet, Jean-Pierre. "Baily, Francis." In *The Biographical Encyclopedia of Astronomers*, edited by Thomas Hockey, Virginia Trimble, Thomas R. Williams, Katherine Bracher, Richard A. Jarrell, Jordan D. Marché, F. Jamil Ragep, JoAnn Palmeri, and Marvin Bolt, 84–85. New York: Springer, 2007.

Lyell, Charles. *Principles of Geology, Being an Attempt to Explain the Former Changes of the Earth's Surface, by Reference to Causes Now in Operation*. Vol. 1. London: John Murry, 1830.

Morrell, Jack, and Arnold Thackray. *Gentlemen of Science: Early Years of the British Association for the Advancement of Science*. Oxford: Oxford University Press, 1982.

Musselman, Elizabeth Green. *Nervous Conditions: Science and the Body Politic in Early Industrial Britain*. Albany: State University of New York Press, 2006.

Newton, Isaac. *The Mathematical Principles of Natural Philosophy*. Translated by Andrew Motte. New York: Daniel Adee, 1846.

Panter, John Raspin. "Sir John F. W. Herschel and Scientific Thought in Early Nineteenth-Century Britain." PhD diss., University of New South Wales, 1976.

Peckham, Morse. "Dr. Lardner's 'Cabinet Cyclopedia.'" *Papers of the Bibliographical Society of America* 45, no. 1 (1951): 37–58.

Pence, Charles H. "Sir John F. W. Herschel and Charles Darwin: Nineteenth-Century Science and Its Methodology." *HOPOS: The Journal of the International Society for the History and Philosophy of Science*, 2018: 108–40.

Playfair, John. Review of *Traité de mécanique céleste*. *Edinburgh Review* 11, no. 22 (1808): 249–84.

Pyle, David M. "Visions of Volcanoes." *Interdisciplinary Studies in the Long Nineteenth Century* 25 (2007): 1–30.

Reidy, Michael S. "John Tyndall's Vertical Physics: From Rock Quarries to Icy Peaks." *Physics in Perspective*, no. 12 (2010): 122–45.

"Report of the Council of the Society to the First Annual General Meeting." *Memoirs of the Royal Astronomical Society* 1 (1822): 21–32.

Review of Herschel's *Discourse on the Study of Natural Philosophy*. *Wesleyan Methodist Magazine* 10 (November 1831): 762–68; continued (December 1831): 842–49.

Richards, Joan L. *Generations of Reason: A Family's Search for Meaning in Post-Newtonian England*. New Haven, CT: Yale University Press, 2021.

Richards, Joan L. "Rigor and Clarity: Foundations of Mathematics in France and England, 1800–1840." *Science in Context* 4, no. 2 (1991): 297–319.

Ross, Sydney. *Nineteenth-Century Attitudes: Men of Science*. Boston: Kluwer Academic, 1991.

Rudwick, Martin J. S. *Bursting the Limits of Time: The Reconstruction of Geohistory in the Age of Revolution*. Chicago: University of Chicago Press, 2005.

Rudwick, Martin J. S. *Worlds before Adam: The Reconstruction of Geohistory in the Age of Reform*. Chicago: University of Chicago Press, 2010.

Ruse, Michael. "Darwin's Debt to Philosophy: An Examination of the Influence of the Philosophical Ideas of John F. W. Herschel and William Whewell on the Development of Charles Darwin's Theory of Evolution." *Studies in the History and Philosophy of Science* 6 (1975): 159–81.

Ruskin, Steven. *John Herschel's Cape Voyage: Private Science, Public Imagination and the Ambitions of Empire.* Aldershot, UK: Ashgate, 2004.

Schaaf, Larry. *Tracings of Light: Sir John Herschel and the Camera Lucida: Drawings from the Graham Nash Collection.* San Francisco: Friends of Photography, 1989.

Schaffer, Simon. "Scientific Discoveries and the End of Natural Philosophy." *Social Studies of Science* 16, no. 3 (1986): 387–420.

Schweber, S. S., ed. *Aspects of the Life and Thought of Sir John Frederick Herschel.* New York: Arno Press, 1981.

Schweber, S. S. "John Herschel and Charles Darwin: A Study in Parallel Lives." *Journal of the History of Biology* 22 (1989): 1–71.

Scott, Walter. *History of Scotland.* London: Longman, Rees, Orme, Brown, and Green, 1830.

Scrope, G. Poulett. *Considerations on Volcanoes: The Probable Causes of their Phenomena, the Laws Which Determine their March, the Disposition of their Products, and their Connexion with the Present State and Past History of the Globe: Leading to an Establishment of a New Theory of the Earth.* London: Phillips, 1825.

Secord, James A. *Victorian Sensation: The Extraordinary Publication, Reception, and Secret Authorship of "Vestiges of the Natural History of Creation."* Chicago: University of Chicago Press, 2000.

Secord, James A. *Visions of Science: Books and Readers at the Dawn of the Victorian Age.* Chicago: University of Chicago Press, 2014.

Shorland, Eileen. *Sir John F. W. Herschel: The Forgotten Philosopher.* The Herschel Family Archive, 2016.

Snyder, Laura J. *The Philosophical Breakfast Club.* New York: Broadway, 2011.

Somerville, Mrs. [Mary]. *Mechanism of the Heavens.* London: John Murray, 1831.

Spence, William. *An Essay on the Theory of the Various Orders of Logarithmic Transcendents: with an Inquiry into Their Applications to the Integral Calculus and the Summation of Series.* London: John Murray, 1809.

Spence, William. *Mathematical Essays of the Late William Spence, Esq. Edited by John F. W. Herschel, Esq. with a Biographical Sketch of the Author.* London: J. Moyes, 1819.

Steinicke, Wolfgang. *Observing and Cataloguing Nebulae and Star Clusters: From Herschel to Dreyer's New General Catalogue.* Cambridge: Cambridge University Press, 2010.

Szabados, László. "von Zach, János Ferenc." In *The Biographical Encyclopedia of*

Astronomers, edited by Thomas Hockey, Virginia Trimble, Thomas R. Williams, Katherine Bracher, Richard A. Jarrell, Jordan D. Marché, JoAnn Palmeri, and Daniel W. E. Green, 2269–70. New York: Springer, 2014.

Taylor, E. G. R. *The Mathematical Practitioners of Hanoverian England, 1714–1840*. Cambridge: Cambridge University Press, 1966.

"Tennant, Smithson." In *Complete Dictionary of Scientific Biography*, vol. 13, 280–81. Detroit: Charles Scribner's Sons, 2008.

Todd, A. C. *Beyond the Blaze: A Biography of Davies Gilbert*. Truro, Cornwall: D. Bradford Barton, 1967.

Topham, Jonathan R. "Science, Print, and Crossing Borders: Importing French Science Books into Britain, 1789–1815." In *Geographies of Nineteenth-Century Science*, edited by David N. Livingstone and Charles W. J. Withers, 311–44. Chicago: University of Chicago Press, 2011.

Turner, Frank M. "The Victorian Crisis of Faith and the Faith that was Lost." In *Victorian Faith in Crisis: Essays on Continuity and Change in Nineteenth-Century Religious Belief*, edited by Richard J. Helmstadter and Bernard V. Lightman, 9–38. Stanford, CA: Stanford University Press, 1990.

Tyndall, John. *The Forms of Water in Clouds and Rivers, Ice and Glaciers*. New York: D. Appleton, 1897.

Voskulh, Adelheid. "Recreating Herschel's Actinometry: An Essay in the Historiography of Experimental Practice." *British Journal for the History of Science* 30, no. 3 (1997): 337–55.

Warner, Brian, and John Rourke. *Flora Herscheliana: Sir John and Lady Herschel at the Cape 1834 to 1838*. Johannesburg, South Africa: Brenthurst, 1998.

Warwick, Andrew. *Masters of Theory: Cambridge and the Rise of Mathematical Physics*. Chicago: University of Chicago Press, 2003.

Watts, Iain P. "'We Want No Authors': William Nicholson and the Contested Role of the Scientific Journal in Britain, 1797–1813." *British Journal for the History of Science* 47, no. 3 (2014): 397–419.

Whewell, William. Review of *A Preliminary Discourse on the Study of Natural Philosophy* by J. F. W. Herschel. *Quarterly Review* 45 (1831): 374–407.

Winterburn, Emily. "The Herschels: A Scientific Family in Training." PhD diss., Imperial College, 2011.

Winterburn, Emily. *The Quiet Revolution of Caroline Herschel: The Lost Heroine of Astronomy*. Stroud, UK: The History Press, 2017.

Woodhouse, Robert. *A Treatise on Plane and Spherical Trigonometry*. Cambridge: Black, Perry, and Kingsbury, 1809.

"Woodhouse, Robert." In *Complete Dictionary of Scientific Biography*, vol. 14, 500. Detroit, MI: Charles Scribner's Sons, 2008.

Yeo, Richard. *Defining Science: William Whewell, Natural Knowledge, and Pub-

lic Debate in Early Victorian Britain. Cambridge: Cambridge University Press, 1993.

Yeo, Richard. "Reading Encyclopedias: Science and the Organization of Knowledge in British Dictionaries of Arts and Sciences, 1730–1850." *Isis* 82, no. 1 (1991): 24–49.

Yeo, Richard. "Reviewing Herschel's *Discourse.*" *Studies in the History and Philosophy of Science* 20, no. 4 (1989): 541–52.

INDEX

Note: Page numbers in *italics* indicate figures.

aberration, 83, 92, 116, 142

Académie des Sciences, 14, 72, 76, 77–78, 86, 104, 126

acoustics, 221

actinometer, 127

Act of Abolition (1807), 28

Agrigento, Italy, 136

Aiguille de Dieux, Mont Blanc (Herschel), *108*

air, weight of, 221

Airy, George, 115, 175, 229

algebra, 68–69

Algebra (Herschel), 69, 239n57

Alps/Alpine expedition, 97, 117, 126, 133, 141, 144, 220; as laboratory of nature, 97; measurement and experiment in, 101–14; peaks, physical measure of, 105

Amici, Giovanni Battista, 127–29

Amiens, Peace of, 11, 12, 71

Ampère, André-Marie, 126

analysis, 13, 24–25; benefits, 48; geometrical notions, 24; Lacroix (*See* Lacroix, Sylvestre François);

Lagrangian version, 63–64, 65; methods, 25; promotion, 42; reforming, 79–82. *See also* Analytical Society

analytical expression, 29, 48

"Analytical Formula for the Tangent, Cotangent, Etc." (Herschel), 32–33

Analytical Society, 34–43, 84, 98, 213–14, 225; effect/impact, 34–35; failure, 86; final triumph, 149; ideals, 86; international reach, 90; Lacroix translation, 63–65, 68, 69, 73, 77, 82; Laplacian program and, 74; mathematical reforms, 103; meetings, 46, 115; *Memoirs of the Analytical Society,* 42, 43, 46–55, 58, 60, 62, 64, 67–69, 86; Newton's calculus, 49; original members (new), 34; radical program, 37; reforming, 62–69, 72; work of, 42, 58, 190, 194, 210–11

Andromeda Nebula, 158, 159

Anglican faith, 166, 167

Annales de chimie, 193

antirevolutionary riots, 30

Arabian Nights, 139

Arago, François, 77, 78, 103–4, 126, 128, 217

Arcueil, Paris, 14, 71, 72

Astronomical Society, 103, 116, 120, 122, 124, 145, 152, 190; Baily and, 84–90, 95, 203; Board of Longitude and, 97; bulletins, 87; business astronomers, 88, 123; businesslike efficiency, 206, 226; coordinated surveys, 87; foreign astronomers, 87; foreign secretary, 88, 114, 127; formation, 52, 84–86; founders, 34, 104; ideals, 86; international reach, 90; leadership, 88, 95; measuring double stars, 91; members/membership, 88, 95, 127; *Nautical Almanac* and, 163; presidency of, 147, 159, 162; primary purpose, 226; profit, 95; as a reforming/transformative society, 34, 52, 68; Royal Society and, 52, 89, 93, 94, 95, 159–60; Slough catalog and, 229; Zach and, 128

astronomical surveys, 4, 87

astronomy, 21, 40, 52; calculational, 82–84, 87; catalog of nebulae and stars (*See* catalog of nebulae and stars); Christianity and, 18; Cuvier's work and, 15; division of labor, 218; instruments (*See* instruments/instrument makers); John Herschel and, 4, 66–67, 69, 72, 85–95, 97, 208–9, 222–24, 226, 229, 230–31, 232; Laplace's theory and, 15, 18, 25, 26, 78; mathematical, 26, 36, 82–85; Napoleon and, 16–17; Newto-nian gravity and, 36; positional, 77, 83–84, 87, 89, 164; sidereal, 22; techniques and systems, 218; William Herschel and, 3, 4, 11–18, 21–23, 25, 36, 45, 46, 65, 66–67, 208, 220, 229. *See also* Astronomical Society; Royal Society

atomic theory, 223

Auvergne, 168–70

Babbage, Charles, 9, 59, 60, 61–68, 70, 96, 97, 115, 124–25, 128, 138, 141, 145, 169, 176, 189, 193, 197–98, 205–6, 211; Alpine expedition, 97, 101–14, 117, 126, 133; as an accountant, 170; Analytical Society, 33–34, 35, 37, 42; behavior and personality trait, 19, 20, 51, 95; calculating machine/difference engine, 125, 170–73, 187–88, 192, 195; Cambridge and, 19, 20, 30, 31, 33–35, 37–38, 42, 58, 172–73, 175; domestic life, 170, 171–72; as father of computing, 9; Lacroix's standard textbook on calculus, 30, 31, 63; London and, 58; Lucasian Chair, 175; marriage, 96, 170, 172; *Memoirs of the Analytical Society*, 42, 43, 46–55, 58, 60, 62, 64, 67–69, 86; on neglected state of science, 192; Paris trip, 72–79; personal losses, 172, 175; *Reflections on the Decline of Science*, 194–96, 197, 201, 202, 204; relationship with his father, 170; Royal Society and, 94, 95, 152–56, 161, 162, 201; Senate House exams, 20; solo trip through France and into Italy, 172; as trustee of Herschel and Harriet's marriage settlement, 99–101

Babbage, Georgiana, 96, 100, 154, 168, 170, 172

Bacon, Francis, 7, 210, 216

Baily, Francis, 82–90, 93, 95, 116, 124, 152, 154, 160, 203, 204, 205, 226

Banks, Joseph, 36, 73, 82–83, 92, 94, 95

barometer, 105, 110, 112, 113, 127, 129–30, 141, 151, 221

Beagle, 208, 228

Berlin Ephemeris (Encke), 164

Berthollet, Claude Louis, 14, 71, 72

Berzelius, Jacob, 77

Bessel, Friedrich, 90

Bible Society, 33, 51, 63

Biot, Jean-Baptiste, 75, 76–81, 103–4, 126, 128, 149

Board of Longitude, Britain, 85, 97, 123, 124, 146, 151, 204; dissolution, 162, 163–64, 174, 177, 192; establishing, 78; Glass Committee, 147, 150, 162, 163; members, 78; *Nautical Almanac*, 83, 84, 87, 116, 123, 162, 163, 171, 197; prizes, 83; resident commissioner, 116, 147

Bode, Johann Elert, 90

Bologna, 129, 140

Bonaparte, Napoleon, 43, 105, 135, 169; astronomy and, 16–17; autocracy, 18; Berthollet and, 71; centralizing programs, 74; Cuvier and, 15; defeat at Waterloo in 1815, 24, 72; fall from power, 43, 72, 74, 77; Laplace and, 12, 17, 18, 167; military campaigns/wars, 11, 24, 30; science and, 13, 16; William Herschel and, 16–17

Boone, Daniel, 82

Boulton, Matthew, 186

Bouvard, Alexis, 90

Brewster, David, 25, 69, 70, 73–75, 149, 151, 192

Brioschi, Carlo, 140

Britain: assassination of Perceval, 44–45, 55; economic disruptions, 41; French war, 11, 24; as a global superpower, 45; learned societies, 46–47; literacy rates, 189; mathematical situation, 25; middle class, 189; science in, 14, 188. *See also* Astronomical Society; Cambridge University; London; Royal Society

British Association for the Advancement of Science, 227

Bromhead, Edward, 34, 37, 38–39, 50, 51, 53, 62, 63

Buckland, William, 102, 125, 134, 139, 169

Bureau des Longitudes, 72, 78, 103

Cabinet Cyclopedia (Lardner), 190, 196, 197, 209, 210; introductory prefaces, 196

Cagnoli, Antonio, 84

Calais, 74, 117, 125

calculational astronomy, 82–84, 87

calculus, 13, 25, 49; abstract foundation, 25; Lacroix, 63–65; Lagrange, 63–64. *See also* mathematics

Cambridge University: Analytical Society (*See* Analytical Society); Babbage and, 19, 20, 46, 47, 51, 58, 63; British Association for the Advancement of Science meeting, 227, 232; curricular changes, 211; Darwin and, 209; Dealtry and, 3, 10; fellowships

(*See* fellowships); John Herschel and, 3, 4, 6, 9, 10, 19–20, 23, 25, 26, 27–43, 45, 46, 51, 57, 59–62; law education, 53, 59; Lucasian Chair of Mathematics, 114–15, 173–74, 175; mathematical exams/tests (*See* mathematical exams/tests); mathematical reforms, 29–30, 31–37, 62–69, 210–11, 225; mineralogy, 57; Newton and, 21; observatory at, 173; Plumian professorship, 174–75; Pollock on, 19; professorship, 59–60; St John's College, 9, *27*, 27–28, 32, 33, 35, 51, 56, 59, 132, 173; students entering, 20–21; Trinity College, 19, 26, 27, 33, 85; Woodhouse and, 19, 29–30, 32–33, 47, 104

camera lucida, 105–6, 110, 113, 117, 130, 151, 186; *Aiguille de Dieux, Mont Blanc, 108*; described, 97; *Etna-the summit called the "Bicorne" from Casa Inglese, 137*; *Goderick Castle, Monmouth, 187*; *Interior of the Crater Vesuvius, 132*; *Interior of the Roman Colosseum, 131*; *Netley Abbey. Southampton. East Window & South Transept, 178*; *Sasso Vernale Valley of Fossa Tyrol from the Gorge of Fredaja, 140*; *Vale of Suza, being the entrance into Piedmont descending from Mont Cenis, 109*; *Vesuvius from the Chiaja St. Lucia Naples, 120*; *View from the Hotel Window, Leamington, Warwick, 186*

Cape of Good Hope, South Africa, 8, 9, 95, 125, 146, 170, 208–9, 228, 230–32

Cape Town, South Africa, 208

Carlini, Francesco, 110

catalog of nebulae and stars, 8, 83–84, 86, 87, 220, 222, 226; British observers and, 95, 163, 164; Caroline Herschel's, 21–22, 143, 159; Charles Messier's, 14; Francis Baily's, 83, 203; John Herschel's, 116, 158–59, 168, 208–9, 229, 230; Palermo observatory, 134; Parramatta observatory, 163; Slough catalog, 158, 159, 229, 230; William Herschel's, 21–22, 65, 122, 158, 159

causes, 216–17

Chaîne des Puys, 168

Chamonix, 107, 109–10, 111, 125, 215

Chaptal, Jean-Antoine, 16

Charles II, 35

chemistry, 57–58, 75, 223; experimentation, 60; growth of, 57; minerals and, 57

Child, James, 125

Children, John George, 152

Christianity, 18, 166, 167–68, 170, 184

Christie's auction, 98, 99

Church of England, 32, 33, 52

Clarke, Edward Daniel, 57, 59, 105, 115

Clarkson, Thomas, 28

clergymen, 32

Colebrooke, Henry Thomas, 86

Colossus of the Apennines, 130

Configliachi, Pietro, 110

Considerations on Volcanoes (Scrope), 169

Conybeare, William, 102, 125

Cook, James, 82

Copley Medal, 59, 97, 114

Correspondence Astronomique, 128

Cotes, Roger, 36

Couttet, Joseph-Marie, 107, 111–12, 113

Covelli, Nicola, 133

creatures of reason, 5, 216–20

crystallography, 221; polarized light, 222

Cumming, James, 60

Cuvier, Georges, 15–16, 77, 134, 168, 196

d'Alembert, Jean le Rond, 13

Darwin, Charles, 6, 208, 224, 228; biological revolution, 208, 227; *Origin of Species,* 208, 224

Daubeny, Charles Giles Bridle, 139, 140, 169

Davy, Humphry, 59, 124, 131, 133, 146–47, 171; arrogance and divisiveness, 93; autocratic style, 154; discovery of iodine, 57; resignation, 161; Royal Institution and, 57, 93; Royal Society presidency and, 93–95, 123, 147, 151–54, 157, 161; voltaic pile of, 57

Davy, John, 59

Dawlish, England, 66–67

Dealtry, William, 9–10; *Principles of Fluxions,* 3, 4, 9

Delambre, Jean Baptiste Joseph, 90

Description of Active and Extinct Volcanoes (Daubeny), 169

Dickens, Charles, 133

diffraction, 104, 149

division of labor, 218

Dollond, George, 100, 116, 127, 128, 147, 150, 205

Dollond, Peter, 91

double refraction, 76, 79, 80–81, 104, 217, 240n8

double stars, 175, 222; catalogs, 122,

159, 168; John Herschel's observations/work, 91, 104, 113–14, 116, 122, 128, 158, 159, 160, 168, 169, 218, 226, 230; measuring, 91; Newton's law of gravity and, 65; parallax among, 160; velocity, 169; William Herschel's observations/work, 12, 21–22, 65, 66, 84, 91, 122, 143, 159

École Polytechnique, 74

Edinburgh Philosophical Review, 73

Edinburgh Review, 38

Edinburgh University, 26

electromagnetism, 77, 126, 209

Elliot-Murray-Kynynmound, Gilbert, 106, 113

Emerson, William, 49

Encke, Johann Franz, 142, 164

Encyclopedia Metropolitana, 148, 189–93, 221

encyclopedias, 189–90

equations of differences, 47, 48

equatorial telescopes, 91, 160, 177

Essai sur la géographie des plantes (Humboldt), 105

Etna, 121, 130, 134, 136–38, *137,* 144, 169, 220, 226

Etna–the summit called the "Bicorne" from Casa Inglese (Herschel), *137*

Eton College, 26

Euler, Leonhard, 13

Faraday, Michael, 116, 147, 150, 209–10

Fellenberg, Philipp Emanuel von, 113

fellowships, 32, 41–42, 153, 173

Fitton, William Henry, 202–3, 204, 205

FitzRoy, Robert, 208, 209, 228

Florence, 130

fluxions, 23–24, 49; concept, 24; dot in the notation of, 34
Forbes, James David, 107
fossils: Cuvier's work on, 15, 16; Lamarck's work on, 16
Fourier, Joseph, 77
France, 4, 11–18, 38, 41, 44, 64, 110; Académie des Sciences, 14, 72, 76, 77–78, 86, 104, 126; Auvergne region, 139, 168–70; Bourbon monarchy, 13, 72; geology, 103–5, 139, 168–70; Grahame in, 177, 179, 181; John Herschel's trips/travels, 70, 79, 81, 82, 85, 88, 103–5, 119, 143, 150, 172, 188, 222; Institut de France, 14, 15, 16, 18, 26, 72, 74, 76, 77; introductory treatises, 196; liberal revolutionary ideals, 30; mathematics/mathematical analysis, 10, 18, 24–25, 30, 62, 64, 82; natural philosophers in, 78; revolution (*See* French Revolution); science in, 14, 16, 72, 74, 76, 77–78, 86, 104, 126, 172; war(s) and, 24, 25, 30–31, 41, 45, 70, 71. *See also* Bonaparte, Napoleon; Laplace, Pierre Simon; Paris, France
Fraunhofer, Bavarian Joseph von, 127, 140, 141–42, 144, 150, 215
Frederick, Augustus (Duke of Sussex), 202, 203, 209, 227, 231
French Revolution, 11, 12, 13, 72, 166
Fresnel, Augustin-Jean, 104

Galilei, Galileo, 7
Gardiner, Marguerite, 139–40
Gauss, Friedrich, 90
Gay-Lussac, Joseph Louis, 77
Geneva, 106–7

Genoa, 128, 140
Geological Society, 59, 89, 102, 125, 169, 202, 206
geology, 8, 18, 36, 117, 125, 139, 144, 172, 231; Auvergne, 139, 168–70; Britain, 102–3, 105; cataclysmic changes/events, 16, 102–3; as a cosmical phenomenon, 222; Dawlish, 66; debates, 15, 18; Earth history, 102–3, 222; as a field of study, 102; FitzRoy and, 208, 209; French, 103–5, 139, 168–69; Geneva, 106–7; illustrations from, 226; inductive science, 222; Italian, 103, 126–38, 141, 144, 158; laboratory, 222; Lamarck's work, 15–16; mineralogy and, 223; Mont Blanc, 106–8, *108*, 109, 111, 112, 215; observations on, 87, 208; Smith's map of England and Wales, 102; theory of Earth, 87; volcanoes, 105, 119, 121, 126, 131–33, *134*, 137–39, 168–70, 176, 221. *See also* Alps/Alpine expedition
geometry/geometrical representation, 36–37, 49
George I, 143
George III, 12, 45, 88, 206
Giambologna, 130
Gilbert, Davies, 146–47, 152, 156, 157, 160–62, 171, 202–7, 227
Glass Committee, 147, 150, 162, 163
Goderick Castle, Monmouth (Herschel), *187*
Gompertz, Benjamin, 86
goniometer, 81, 105, 223
Gordon, George (Lord Byron), 119, 121, 140
Grahame, James, 9, *39*, 238n39; as an amateur historian, 117; Bab-

bage and, 156; bachelorhood, 176; family, 40; financial difficulties, 177; Herschel and, 37, 38, 40–41, 42, 53, 55, 56, 59, 96, 117–18, 156, 167, 168, 176–83, 186, 188–89, 225; illness, 117; law career and, 40, 53–54; legal/law studies for, 55, 56, 59; London, 176, 177, 178, 188; mathematics and, 41; Nantes, France, 177, 179, 181; plans for Herschel's marital happiness, 177; religion and, 167; Stewarts and, 177, 178, 179, 182; tracts on abolition of slavery, 40; Wesleys and, 177, 179

Grahame, Matilda, 176, 178
Grahame, Robert, 40, 53
Gray, Thomas, 165
Greenough, George Bellas, 125
Greenwich Observatory, Britain, 122, 123
Gretton, George, 26
Gurney, Hudson, 197
Gwatkin, Harriet, 97, 98–99
Gwatkin, Richard, 34, 63, 98
Gwatkin, Robert Lovell, 98, 99
Gwatkin, Theophila, 98

Halley, Edmond, 35
Hannover, 71, 143
Harding, Karl Ludwig, 90
Haüy, René Just, 14
Herschel, Alexander, 22, 23, 67
Herschel, Caroline, 11, 156, 158, 176, 184, 188, 195; catalogs of nebulae and stars, 21–22, 143, 158; home in Hannover, 65, 142, 143; John Herschel's childhood and, 23–24; John Herschel's letter to, 138; research/observations/work, 21, 22, 65, 142–43, 158, 229; sidereal astronomy, 22, 66, 121–22, 143; William Herschel's death and, 143
Herschel, Constance, 20
Herschel, John, 3; accident (overturned carriage), 129–30, 167; *Algebra,* 69, 239n57; Alpine expedition (*See* Alps/Alpine expedition); Analytical Society (*See* Analytical Society); as an author, 189–90; approach/ideas to scientific pursuits, 173–76; artistic tastes, 98; Astronomical Society and (*See* Astronomical Society); Babbage and (*See* Babbage, Charles); baptism, 166; as baronet, 231; childhood, 18, 22–23, 71, 165–66; Copley Medal, 59, 97, 114; correspondence network, 90; education, 19–20, 25–42; engagement (broken) with Harriet Gwatkin, 97, 98–101, 103, 178; European grand tour, 119–21; family wealth, 45–46; fellowship, 51, 67, 173; financial independence, 119; in-home tutors, 26; Lacroix translation, 62–65, 68, 69, 73, 82; love letter to Margaret, 183; marriage/wedding, 185; mathematical training, 25–26, 31; *Memoirs of the Analytical Society,* 42, 43, 46–55, 58, 60, 62, 64, 67–69, 86; model of theory formation, 227; natural philosophy, 4–5; Paris, trip to, 10, 11, 13, 38, 70, 72–79, 80, 81, 97, 103–6, 110, 113, 125–27, 128, 144, 148, 168; poetry and poetic sense, 130; professorship, 173, 174–75; religious faith, 166, 183–84; Reynolds artwork/drawings and,

98, 99, 101; Royal Society and (*See* Royal Society); as a scholar, 28; scientific credibility, 170; scientific virtues, 212, 213; Senate House exams, 20; as senior wrangler, 20; social life, 92, 160; Stewarts and, 178–84

Herschel, Mary Pitt, 11, 22, 25, 26, 99, 101, 117

Herschel, William, 4, 10, 65–67; an immigrant to England, 11–12; astronomy and, 3, 4, 11–18, 21–23, 25, 36, 45, 46, 65, 66–67, 208, 220, 229; catalog of nebulae and stars, 21–22, 65, 122, 158, 159; death, 17, 97–98, 117–18, 119, 168; discovery of a new world, 11, 12; health, 65, 67; marriage, 22; as a musician, 11–12; polished surfaces, 46; Robert Gwatkin meeting, 99; telescopes, 4, 12, 23, 25, 36, 38, 39, 43, 46, 57, 65, 66, 67, 88, 90, 91; telescopic surveys, 18; Wilson and, 26

Hitcham House, 26

Hooke, Robert, 35, 157

hope, scientific virtue of, 212, 213

Humboldt, Alexander von, 105, 126

humility, scientific virtue of, 212, 213

Hutton, James, 16

hydrostatics, 221

hyposulphurous acid, 73, 79

Iceland spar, 75–76

Illustrations of the Huttonian Theory of the Earth (Playfair), 16

inductive method, 5

Industrial Revolution, 38

Institut de France, 14, 15, 16, 18, 26, 72, 74, 76, 77

instruments/instrument makers, 86, 100, 116, 118, 148, 205; aberration, 92; Alpine peaks and landscape measurement, 105–6, 112, 113; Board of Longitude, 151, 162; British, 87–88, 91, 92, 120–21, 127, 151, 156; European, 120, 127–29; fuzzy objects measurements, 159; John Herschel's, 66, 91–92, 105–6, 125, 126–27, 129–30; Italian observatory, 144; naval vessels, 151; observatories and, 126, 127, 129, 144; Reichenbach's transit circles, 141; William Herschel's, 21, 23, 45, 46, 66. *See also specific instruments*

Interior of the Crater Vesuvius, June 9, 1824 (Herschel), *132*

Interior of the Roman Colosseum, May 18, 1824 (Herschel), *131*

intermolecular forces, 223

inverse cosine function, 37

Italy, 158, 168, 172, 193, 222; accident (overturned carriage), 129–30, 167; Alps (*See* Alps/Alpine expedition); astronomers, 97, 110, 127–29, 182; Cagnoli's work, 84; French invasion of, 11; geological and astronomical observations, 126–38, 141, 144, 158; historic sights, 125; Lord Byron traveling to, 119, 121; opticians and instrument producers, 120; Romantic ideal, 119, 121; Vesuvius (*See* Vesuvius); volcanoes of, 119, 121, 126, 131–38, 168, 231

James II of England (James VII of Scotland), 38

Joint Glass Committee. *See* Glass Committee

Jones, Richard, 52, 96, 103, 117, 198–99
Jorio, Andrea de, 134
Journal of Natural History, Chemistry, and the Arts, 32

Kelly, Patrick, 86

Lacroix, Sylvestre François, 14, 25, 27, 30, 33, 62–65, 68, 69, 73, 77, 82. *See also Traité élémentaire du calcul différentiel et du calcul intégral* (Lacroix)
Lagrange, Joseph-Louis, 13, 27, 63–64; calculus, 63–64; *Mécanique analytique,* 24, 26
Lamarck, Jean-Baptiste, 15–16
Laplace, Pierre Simon, 12–13, 18; Berthollet and, 71; disciples, 13; *Mécanique céleste,* 13, 17, 25, 26, 38, 47, 74, 197; political savvy, 77; scientific research program, 74; on William Herschel's work, 104
Lardner, Dionysius, 190–92, 196, 197, 200–201, 205, 207, 208, 209, 210, 211, 250n37
Lavoisier, Antoine, 12, 57
laws, 225; abortive study of, 225; abstract, 219; governing society, 225; nature, 219–20; quantitative, 219; scientific, 217; universal application, 219
lawyers, 53–55
Lee, Stephen, 124
Legendre, Adrien-Marie, 77
light, 148–50, 189–91; directional property, 76; Huygensian theory of, 221; Iceland spar, 75–76; mathematical expression, 149; nature of, 104; Newtonian theory of, 221; polarization, 75,

80, 80–81, 148, 149, 191, 217, 221–22; synthetic treatment of, 148
"Light" (Herschel's treatise), 148, 150, 157, 164, 189–90
Lincoln's Inn, 56
logic, 213
London, 6, 22, 34, 44, 54, 73–74, 76, 79, 91–95, 103, 109, 112, 116–18, 176–79, 180; astronomical instruments/instrument makers, 87–88, 91, 127; chemical experimentation, 58–59; Christie's auction, 98, 99; as commercial powerhouse, 56; lawyers, 53; middle class, 52, 86; physical discoveries, 57–58; population, 55; professional class, 68, 72; publishing industry, 5; Reynolds artwork/drawings and, 98, 99, 101; scientific community, 70, 73, 89, 114, 116, 120, 209, 211, 227–28; scientific elite, 83, 144; scientific institutions, 145–64, 170; as warehouse of the world, 55–56. *See also* Astronomical Society; Geological Society; Royal Society
Longmans, 190
Lubenow, William, 7
Lucasian Chair of Mathematics, 114–15, 173–74, 175
lunar occultations, 84, 87, 116
Lyell, Charles, 131, 134, 152

Mackintosh, James, 210
Maclear, Thomas, 228
Malus, Étienne-Louis, 75–76
marriage(s)/marriage settlement, 98–101; trustees, 99–100
Maskelyne, Nevil, 28, 122
mathematical exams/tests, 28–29;

Senate House exams, 20, 28, 37, 41–43, 51, 64; Smith's Prize, 29, 42; Tripos, 28–29, 30, 41

mathematics, 3, 45, 213; advanced, 31; algebraic operations, 37; extracurricular work, 48; inverse cosine function, 37; reforms, 29–30, 31–37, 62–69, 210–11, 225; Rogers on, 31; spatial reasoning, 49; trigonometric functions, 37. *See also* analysis; astronomy

matter: division of, 221; polarity of, 221

Mécanique analytique (Lagrange), 24, 26

Mécanique céleste (Laplace), 13, 17, 25, 26, 38, 47, 74, 197

Mechanism of the Heavens (Somerville), 197

Memoirs of the Analytical Society, 42, 43, 46–55, 58, 60, 62, 64, 67–69, 86

Mer de Glace glacier, 107–8, 112

Messier, Charles, 14

Methodist movement, 166

Milky Way, 118, 121, 135, 138

Mill, John Stuart, 5, 212

Milner, Isaac, 114–15

Milton, John, 130

mineralogy, 81, 223

mirrors, 67, 69

Mitscherlich, Eilhard, 126

Modena, 128–29

Monatliche Correspondenz, 128

Montanvert. *See* Mer de Glace glacier

Mont Blanc, 106–8, *108*, 109, 111, 112, 215

Mont Cenis, *109*, 109–10

Monte Cuccio, 136

Monte Rosa, 111–13, 126

Monticelli, Teodoro, 131

Mount Etna. *See* Etna

Naples, 106, 119, *120*, 121, 130–31, 133–35, 139–40, 172, 223

natural philosophy, 7–8; reforming, 212–16; religious aspect, 212–13; as a tool for virtue, 227. *See also* astronomy; mathematics; science

natural selection, 208

Natural Theology, or Evidences of the Existence and Attributes of the Deity (Paley), 166

nature: knowledge of, 215; laws, 214; sublime legislation, 214

Nautical Almanac, 83, 84, 87, 116, 123, 162, 163, 171, 197

nebular hypothesis, 15. *See also* catalog of nebulae and stars

Netley Abbey. Southampton. East Window & South Transept (Herschel), *178*

Newton, Isaac, 3–4, 13, 21, 35; corpuscular theory, 75; fluxions, 23–24; gravitational/gravity theory, 149; physical properties, 24; *Principia*, 20, 23, 166; sound induction, 221

Nicholson, William, 32, 35, 47

notation, 48–49

O'Brien, Mary, 98

observational astronomy, 67, 83–84

"On a Remarkable Application of Cotes's Theorem" (Herschel), 36

optics, 148–50, 189–90, 222; glass properties, 148–49; laws, 149. *See also* light

Oriani, Barnaba, 90

Origin of Species (Darwin), 208, 224

Orion Nebula, 158, 159

paleontology, 15

Palermo, 134–36, 138–39, 167

Paley, William, 166

Paradise Lost (Milton), 130

Paris, France, 73–79; John Herschel's trip to, 10, 11, 13, 38, 70, 72–79, 80, 81, 97, 103–6, 110, 113, 125–27, 128, 144, 148, 168; longitude between Greenwich and, 151, 162; Peace of Amiens, 11; scientific monopoly of, 74; as social and political center, 74; William Herschel's trip or visit to, 10, 11–18, 24, 26, 71

Parramatta Observatory in Australia, 163–64

Pascal, Blaise, 221

Peacock, George, 34, 63, 64–65, 68, 69, 85, 94, 103, 105, 173, 211

Pearson, William, 85–86, 88, 104

Peel, Robert, 160–62, 205

Perceval, Spencer, 44–46, 55

Périer, Florin, 221

Philosophical Transactions of the Royal Society, 29, 32, 35–36, 47, 49, 58, 61, 63, 68, 73, 84, 102, 114, 146, 160, 193, 229

photography, 73, 106

"Physical Astronomy" (Herschel), 190

physical sciences, 213

Piazzi, Giuseppe, 135–36

Plana, Giovanni, 110, 127–28, 182

Playfair, John, 16, 25, 38, 41, 62

Plumian Chair of Astronomy, 115

pneumatics, 221

poetry and poetic sense, 130

Poisson, Siméon Denis, 77

polarity, 221

polarized light, 75, *80,* 80–81, 148, 149, 191, 217, 221–22

Pollock, Frederick, 19

Pompeii, 131–32, 133

Pond, John, 122–24

Pons, Jean-Louis, 128

positional astronomy, 77, 83–84, 89, 164

Pozzuoli, 133–34

Preliminary Discourse in the Study of Natural Philosophy, A (Herschel), 5–9, 207, 209–13, 215, 217, 223, 232; claims of, 201; drafting/writing, 196–201; English religious thought, 211; model for theories, 224; as a philosophy of physical science, 209; publication, 209; reviews, 209–10; revisions to, 202, 207; as a vehicle of scientific reform, 210; vision of science, 224–28

Priestly, Joseph, 30, 57

Principia (Newton), 20, 23, 166

Principles of Fluxions (Dealtry), 3, 4

printing technology, 49–50

reasoning, 213, 216–20; classification, 218; inductive, 218–19; mathematical, 213–14; observations, 218; phenomenon and, 219; theories and, 220

reflecting telescopes, 97; Amici creating, 127, 128; refracting *vs.*, 91–92; William Herschel's, 4, 12, 23, 25, 36, 38, 39, 43, 46, 57, 65, 66, 67, 88, 90, 91

Reflections on the Decline of Science (Babbage), 194–96, 197, 201, 202, 204

reflective goniometer. *See* goniometer

refracting telescopes, 67, 97, 116; Fraunhofer, 140; high-quality glass, 142; improvement, 92; reflecting *vs.*, 91–92; South, 116

Regency London, 3

Reichenbach, Georg Friedrich von, 141

Reid, Thomas, 28

Reynolds, Joshua, 98, 99, 101, 241n2

Rogers, Alexander, 26, 29, 31, 52

Rome, 130, 131

Royal Academy of Naples, 131

Royal Botanic Gardens, 82

Royal Institution, 56–59, 89, 93

Royal Observatory at Greenwich, 83

Royal Society, 6, 14, 47, 52, 56–60, 72, 76, 78, 81, 82–86, 89, 97, 103, 110, 116, 123, 124, 145–47, 171, 172, 175, 176, 184; aristocratic qualities and, 203; controversies within, 175; Copley Medal, 59, 97, 114; cronyism and favoritism, 153; government and, 14; membership, 56; parliamentary corruption, 202; *Philosophical Transactions*, 29, 32, 35–36, 47, 49, 58, 61, 63, 68, 73, 84, 102, 114, 146, 160, 193, 229; presidential election, 93–95, 202–7; private dealings, 202; reforming, 150–64, 192–207; as rotten borough, 202–3; as a status symbol, 56; weekly meetings, 56

Royal Society Council, 95

Ruskin, John, 133

Sabine, Edward, 151, 162

Saint Laurence Church in Slough, 118, 165–66, 168

Sanskrit, 86

Santini, Giovanni Sante Gaspero, 140–41

Sasso Vernale Valley of Fossa Tyrol from the Gorge of Fredaja (Herschel), *140*

Saussure, Horace Bénédict de, 107, 111

science, 213, 220–27; church living and, 32; as concern, 14; defini-tion of, 213; division of labor, 218; goal of, 214, 217; as plutoc-racy, 7; practice of, 215; rational thought and, 223–24; terms with unclear meanings as errors in, 214; vision of, 207, 225–27. *See also* astronomy; mathematics

Scientific Revolution, 7

scientific virtues: hope and humility, 212, 213

scientist as a term, 31

Scotland, 26, 35, 38, 41, 53, 59, 73, 117, 151, 182, 191; French influence, 25, 38; mathematical practices, 38; royal succession, 38

Scott, Walter, 191

Scrope, George Julius Poulett, 132, 139, 169

Secord, James, 5

Sedgwick, Adam, 115, 176, 205

Selinunte, Italy, 136

Senate House exams, 20, 28, 37, 41–43, 51, 64

Serapis, Pozzuoli, 133–34

sextants, 105, 151

Shelly, Percy Bysshe, 133

Sicily, 121, 130, 134–39, 143, 144, 169, 223, 230

silver halide, 73

Sismondi, Jean Charles Léonard de, 106

slave trade, 28

Slegg, Michael, 33, 34

Slough, England, 12, 36, 37, 61, 157–60; Baldwin family, 22; catalog, 158, 159, 229, 230; domestic base, 189; Gwatkin meeting William Herschel in, 99; holiday, 183; James visiting, 53; John Herschel's education/ life/work at, 22–23, 43, 45–46,

59, 67, 79, 90–92, 116, 121, 122, 135, 144, 148, 162, 173, 176, 177, 179, 204, 217, 229–30; Jones and Whewell stay at, 198; Peacock and Babbage stay at, 85; Piazzi meeting William Herschel in, 135–36; reflectors at, 122; renovations of house at, 195–96; Rogers traveling to, 26; Saint Laurence Church, 118, 165–66, 168; Spence's manuscripts delivered to, 68; telescopes at, 28, 36, 67, 88, 90–92; William Herschel's arrival in, 22; William Herschel's death in, 117–18

Slough catalog, 229

Smith, William, 102

Smith's Prize, 29, 42

Smyth, William Henry, 159–60

Society for the Diffusion of Useful Knowledge, 189

solar eclipses, 87

solar system: as a complex physical system, 149; gravity and, 222; Laplace's theory or explanation, 14–15, 25, 166–67; visible planets, 12. See also astronomy

Soldner, Johann Georg von, 142

Somerville, Mary, 197, 200, 214

"Sound" (Herschel's treatise), 190, 193

sound induction, 221

South, James, 91–92, 95, 100, 105, 116, 122–25, 128, 152, 156, 159–60, 162, 205, 206

speculum, 67

Spence, William, 68, 69, 74, 194

St. John's College, Cambridge, 9, 27, 27–28, 32, 33, 35, 51, 56, 59, 132, 173

St. Marylebone Parish Church, 177, 180, 185

St. Maur, Edward Adolphus (Duke of Somerset), 89

stars: catalogs (See catalog of nebulae and stars); double (See double stars); measuring position of, 83. See also astronomy

Staubbach Falls, 113

Stewart, Alexander, 177

Stewart, Dugald, 28

Stewart, Emilia, 177, 178, 179, 180–83, 185

Stewart, Isabella, 182–83, 185, 200, 226

Stewart, Johnnie, 188

Stewart, Margaret Brodie, 9, 40, 176, 178–89, 192, 195–96, 198, 199–200, 205, 211, 226, 228, 230, 231

Stone, John, 230

Struve, Friedrich Georg Wilhelm von, 158

Taylor series, 63

telescope(s), 6, 9, 21, 28, 83, 122; construction and operation, 143; equatorial, 91, 160, 177; Fraunhofer's glass, 141, 142; Italian observatories, 144; John Herschel's, 159, 160, 184, 198, 200, 230; London instrument makers, 87–88; manufacturing, 92; mirrors, 67, 69, 200; optical theory, 221–22; reflecting (See reflecting telescopes); refracting (See refracting telescopes); repairing, 118; William Herschel's, 4, 12, 23, 25, 36, 38, 39, 43, 46, 57, 65, 66, 67, 85, 88, 90, 91, 95, 98, 118, 158–59

Tennant, Smithson, 59–60

Thénard, Jacques, 126

Thénard, Louis Jacques, 77

theology, 166–67

thermometer, 127

Times, 172, 202, 205, 206, 210

Tory, 6, 161, 187

Traité élémentaire du calcul différentiel et du calcul intégral (Lacroix), 33; translation of, 62–65, 68, 69, 73, 82

A Treatise on Astronomy (Herschel), 209

trigonometric series, 47

Trinity College, Cambridge, 19, 26, 27, 33, 85

Tripos, 28–29, 30, 41

Troughton, Edward, 88, 91, 127, 128, 205

trustees, 99–100

Tyndall, John, 96, 107

University of Paris, 75

Uranus, 4, 12, 165, 200

Vale of Suza, being the entrance into Piedmont descending from Mont Cenis (Herschel), *109*

variable stars, 169–70, 218, 231

Vesuvius, 105, 119, *120,* 121, 125, 130, 131–32, *132,* 133, 134, 136–38, 144, 169

Vesuvius from the Chiaja St. Lucia Naples (Herschel), *120*

Victoria, Queen, 231

View from the Hotel Window, Leamington, Warwick, March, 1829 (Herschel), *186*

Vince, Samuel, 115

volcanoes, 105, 119, 121, 126, 131–33, 134, 137–39, 168–70, 176, 221

Vollambrosa, monastery of, 130

Walbeck, Henrik Johan, 90

Warburton, Henry, 161

Warring, Edward, 48–49

Watson, William, 28, 66, 166

Watt, James, 39, 40

weight of air, 221

Wellesley, Arthur (Duke of Wellington), 187–88, 194

Wesley, Charles, 177

Wesley, Charles, Jr., 177, 179, 180, 181–82

Wesley, John, 177

Wesley, Sarah, 177, 179

Wesleyan Methodist Magazine, 210

Whewell, William, 5, 199, 201, 209, 212; Airy and, 115; Analytical Society and, 65; as a bachelor fellow, 176–77; coined the term "scientist," 31; curricular changes and, 211; foreign mathematics, 63; Herschel and, 75, 79, 156, 173–74, 185, 188, 198, 201, 209, 227; Royal Society and, 198, 205

Whigs, 155, 161, 225

Whitehall, 40

Whitmore, Georgiana. *See* Babbage, Georgiana

Whittaker, John, 34

Wilberforce, William, 28

Wilson, Patrick, 26

Wollaston, William Hyde, 57, 81, 93–95, 161–62, 223

Woodhouse, Robert, 19, 29–30, 32–33, 47, 104

Young, Thomas, 25, 38, 75, 116, 151, 163, 197–200

Zach, Franz Xaver von, 128